Integrated Circuits/Microchips

Edited by Kim Ho Yeap
and Jonathan Javier Sayago Hoyos

Published in London, United Kingdom

IntechOpen

Supporting open minds since 2005

Integrated Circuits/Microchips
http://dx.doi.org/10.5772/intechopen.83313
Edited by Kim Ho Yeap and Jonathan Javier Sayago Hoyos

Contributors
Navid Mohammadian, Leszek A. Majewski, Karim Ali, Hao-Li Zhang, Zhi-Ping Fan, Jonathan Sayago, Irina Valitova, Zhihui Yi, Sung Min Park, Daniel Arbet, Viera Stopjakova, Lukas Nagy, Saeed Mian Qaisar, Jiang Cao, Kim Ho Yeap, Siu Hong Loh, Muammar Mohamad Isa

First published in London, United Kingdom, 2020 by IntechOpen
IntechOpen is the global imprint of INTECHOPEN LIMITED, registered in England and Wales, registration number: 11086078, 7th floor, 10 Lower Thames Street, London, EC3R 6AF, United Kingdom
Printed in Croatia

British Library Cataloguing-in-Publication Data
A catalogue record for this book is available from the British Library

Additional hard and PDF copies can be obtained from orders@intechopen.com

Integrated Circuits/Microchips
Edited by Kim Ho Yeap and Jonathan Javier Sayago Hoyos
p. cm.
Print ISBN 978-1-78985-930-0
Online ISBN 978-1-83968-461-6
eBook (PDF) ISBN 978-1-83968-462-3

For EU product safety concerns:
IN TECH d.o.o., Prolaz Marije Krucifikse Kozulić 3, 51000 Rijeka, Croatia, info@intechopen.com or visit our website at intechopen.com.

Meet the editors

Kim Ho Yeap is an Associate Professor at Universiti Tunku Abdul Rahman, Malaysia. He is an IEEE senior member, a Professional Engineer registered with the Board of Engineers, Malaysia, and a Chartered Engineer registered with the UK Engineering Council. He received his BEng (Hons) Electrical and Electronics Engineering from Universiti Teknologi Petronas in 2004, his MSc in microelectronics from Universiti Kebangsaan Malaysia in 2005, and his PhD from Universiti Tunku Abdul Rahman in 2011. In 2008 and 2015, respectively, Dr. Yeap underwent research attachment at the University of Oxford (UK) and Nippon Institute of Technology (Japan). Dr. Yeap is the external examiner and external course assessor of Wawasan Open University. He is also the Editor in Chief of the i-manager's Journal on Digital Signal Processing. He has also been a guest editor for the Journal of Applied Environmental and Biological Sciences and Journal of Fundamental and Applied Sciences. Dr. Yeap has been given the university teaching excellence award, and 20 research grants. He has published more than 100 research articles (including refereed journal papers, conference proceedings, books, and book chapters). When working in Intel corporation, Dr. Yeap was a design engineer in the pre-silicon validation group. He was awarded 4 Kudos awards by Intel for his contributions in the design and verification of the microchip's design for testability (DFT) features.

Jonathan Javier Sayago Hoyos is a research associate at the National Autonomous University of Mexico. During his doctoral studies, Dr. Sayago tackled the problem of achieving low-voltage organic transistors employing electrolytes as the gating medium. His research contributed to shedding light on fundamental physicochemical processes in electrochemical transistors and energy storage devices. After his PhD studies, Dr. Sayago worked as a consultant for Bowhead Health Inc., a Canadian startup company aiming for the commercialization of bioelectronic devices for preventive medical applications. His team designed and built a biosensor device capable of testing 50 μl of blood which led the company to secure a private funding from the world-class Mexican company Grupo Arcoiris. As a Postdoctoral Researcher at the Institute of Renewable Energies, UNAM, Dr. Sayago investigates biocompatible and biodegradable electrodes engineered for energy storage and heat transfer applications. Derived from this work he authored and co-authored over 10 articles in highly recognized international journals, 3 book chapters and participated in numerous international conferences and workshops. Dr. Sayago has taught courses in mathematics, advanced physics laboratory, computer assisted design and 3D printing, and tutoring sessions. He is an active reviewer for the Journal of Power Sources.

Contents

Preface

With the world marching inexorably towards the fourth industrial revolution (IR 4.0), one is now embracing lives with artificial intelligence (AI), the Internet of Things (IoTs), virtual reality (VR) and 5G technology. Wherever we are, whatever we are doing, there are electronic devices that we rely indispensably on. While some of these technologies, such as those fueled with smart, autonomous systems, are seemingly precocious; others have existed for quite a while. These devices range from simple home appliances, entertainment media to complex aeronautical instruments. Clearly, the daily lives of mankind today are interwoven seamlessly with electronics.

Surprising as it may seem, the cornerstone that empowers these electronic devices is nothing more than a mere diminutive semiconductor cube block. More colloquially referred to as the Very-Large-Scale-Integration (VLSI) chip or an integrated circuit (IC) chip or simply a microchip, this semiconductor cube block, approximately the size of a grain of rice, is composed of millions to billions of transistors. The transistors are interconnected in such a way that allows electrical circuitries for certain applications to be realized. Some of these chips serve specific permanent applications and are known as Application Specific Integrated Circuits (ASICS); while others are computing processors that can be programmed for diverse applications. The computer processor, together with its supporting hardware and user interfaces, is known as an embedded system.

In this book, a variety of topics related to microchips are extensively illustrated. The topics encompass the physics of operation of the microchip device, as well as its design methods and applications.

Chapter 1 presents an overview of microchips. In order to allow readers to appreciate the efforts researchers have sacrificed to arrive at the cutting-edge technology that we savor today, the historical development of microchips and its fundamental building block, i.e. the transistor, is first illustrated. This is then followed by a brief explanation of Moore's law – the law that governs the technological progression of microchips. A brief introduction to the field effect transistor – particularly the MOSFET, its operational principle, and the precipitating factors that necessitate the evolution of the planar MOSFETs to the three-dimensional FinFETs is also covered. At the end of the chapter, a walkthrough of the chip fabrication process is succinctly described.

In Chapter 2, an overview of the main challenges and design techniques for ultra-low voltage and low-power analog integrated circuits in nanoscale technologies is illustrated. New design challenges and limitations linked to achieving low voltage operation, low process fluctuation, low device mismatch, and other effects are discussed. In the later part of the chapter, conventional and unconventional design techniques (bulk-driven approach, floating-gate, dynamic threshold, etc.) to design analog integrated circuits towards ultra-low voltage systems and applications are described. Examples of ultra-low voltage analog microchip blocks (such as an operational amplifier, a voltage comparator, a charge pump, etc.) designed in a standard CMOS technology but with an unconventional design approach, are also given.

Chapter 3 describes the recent progresses in the tunnel field effect transistors based on 2-D TMD van-der-Waals heterostructure. The chapter covers the theoretical and computational efforts to understand the working mechanism and the limiting factors in these devices. It also sheds light on the design challenges to be addressed for the development of efficient tunnel field-effect transistors based on 2-D material van-der-Waals heterostructures.

In order to support and promote low-cost and bio-degradable electronics, organic field effect transistors (OFETs) have been introduced. Chapters 4 to 6 present a detailed elaboration on various topics related to OFETs. Since multiple crystalline packing states (crystal polymorphism) exist in the active layer of OFETs, a review on crystal polymorph control is given in Chapter 4. One way to minimize the threshold voltage of an OFET is to reduce the gate dielectric thickness. Chapter 5 discusses some of the most promising strategies towards high capacitance dielectrics for low voltage OFETs. Since OFETs are capable of providing tissue equivalent response to ionizing radiation, Chapter 6 presents the possibility of using different types of OFETs as ionizing and X-ray radiation dosimeters in medical applications.

Chapters 7 to 9 describe some of the recent applications of microchips. Chapter 7 presents several CMOS microchips realized for various optical applications, such as high-definition multimedia interface (HDMI), light detection and ranging (LiDAR), and gigabit Ethernet (GbE). Chapter 8 explains spin-orbit torque magnetic random-access memory (SOT-MRAM) and how it is used to realize reliable, high speed, and energy-efficient on-chip memory. Both non-diode-based SOT-MRAM and diode-based SOT-MRAM cells are discussed in this chapter. The final chapter, Chapter 9, describes the design of a novel offset compensated digital baseline restorer (BLR) and a hybrid interpolator. The behavior of the devices is configured using Very High-Speed Integrated Circuits Hardware Description Language (VHDL) and validated on a Field Programmable Gate Array (FPGA).

Kim Ho Yeap
Associate Professor,
Department of Electronic Engineering,
Faculty of Engineering and Green Technology,
Universiti Tunku Abdul Rahman,
Malaysia

Jonathan Javier Sayago Hoyos
Postdoctoral Researcher,
Instituto de Energías Renovables,
Universidad Nacional Autónoma de México,
México

Section 1

Introduction to Microchips

Chapter 1

Introductory Chapter: Integrated Circuit Chip

Kim Ho Yeap, Muammar Mohamad Isa and Siu Hong Loh

1. Introduction

The technological advancement of integrated circuit chips (or colloquially referred to as an IC, a chip, or a microchip) has progressed in leaps and bounds. In the span of less than half a century, the number of transistors that can be fabricated in a chip and the speed of which have increased close to 500 and 5000 times, respectively. Back in the old days, about five decades ago, the number of transistors found in a chip was, even at its highest count, less than 5000. Take, for example, the first and second commercial microprocessors developed in 1971 and 1972. Fabricated in the large-scale integration (LSI) era, the Intel 4004 4-bit microprocessor comprised merely 2300 transistors and operated with a maximum clock rate of 740 kHz. Similarly, the Intel 8008 8-bit microprocessor released immediately a year later after its 4-bit counterpart comprised merely 3500 transistors in it and operated with a 800 kHz maximum clock rate. Both these two microprocessors were developed using transistors with 10 μm feature size. Today, the number of transistors in a very large-scale integration (VLSI) (or some prefer to call it the giant large-scale integration [GLSI]) chip can possibly reach 10 billion, with a feature size less than 10 nm and a clock rate of about 5 GHz. In April 2019, two of the world's largest semiconductor foundries—Taiwan Semiconductor Manufacturing Company Limited (TSMC) and Samsung Foundry—announced their success in reaching the 5 nm technology node, propelling the miniaturization of transistors one step further to an all new bleeding edge [1]. According to the announcement made in the IEEE International Electron Devices Meeting in San Francisco, the TSMC's 5 nm chip would be produced in high volume in the first half of 2020 [2, 3]. TSMC has also started work on their 3 nm nodes [3].

There is little doubt that the electronics world has experienced a quantum leap in its technology for the past 50 years or so and this, to a large extent, is due to the rapid improvement in the performance, power, area, cost and "time to market" of an IC chip. This chapter provides a succinct illustration on the historical evolution of the IC chip, a general overview of the fundamental building block of the chip—the field-effect transistors, and a brief description of the IC design process.

2. A brief history

The thermionic triode was regarded as the predecessor of transistors that are prevalently used to build electronic devices today. Being invented in 1907, the triodes were made of vacuum tubes which were relatively large in size and were naturally cumbersome to be used. In December 1947, however, three physicists working in the AT&T Bell Laboratories— Dr. Walter Houser Brattain, Dr. John Bardeen and Dr. William Bradford Shockley, Jr.—achieved a remarkable scientific

IntechOpen

breakthrough when they successfully constructed the solid-state equivalence of the thermionic triode, i.e. the first point-contact germanium transistor. As can be seen in **Figure 1**, the solid-state transistor that they developed consisted of an n-type germanium crystal block and two layers of gold foils placed in close proximity with each other. The foils acted as the contacts of the transistor. Together with the contact at the base, the transistor had a total of three contacts—which were named the emitter, collector and base contacts. When a small current was applied to the emitter contact, the output current at the emitter and base contacts would be amplified [4, 5].

In comparison with its predecessor, the solid-state transistor was diminutive in size. It also consumed much lower power, operated at relatively lower temperature and gave significantly faster response time. It is therefore apparent that the semiconductor transistor is more superior to its conventional vacuum tube brethren. Owing to its advantages and viability, the vacuum tubes were eventually replaced by the solid-state electronic devices. The inexorable widespread usage of the semiconductor transistors in electronic circuits has triggered a dramatic revolution in the electronic industries, kicking off the era of semiconductor. Because of this significant contribution, Bardeen, Brattain and Shockley were awarded the Nobel Prize in Physics in 1956 [4].

It is worthwhile noting that, when the solid-state device was first introduced, it was not coined the term "transistor." Instead, it was generally referred to as the "semiconductor triode." According to the "Memorandum for File" published by the Bell Telephone Laboratories (BTL) [6], six names had been proposed for the device—namely, "semiconductor triode," "surface states triode," "crystal triode," "solid triode," "iotatron" and "transistor." The word "transistor" (which originates from the abbreviated combinations of the words "transconductance" and "varistor") proposed by Dr. John Robinson Pierce of BTL had ultimately turned out to be the winner of the internal poll [6].

The first silicon transistor was developed by Dr. Morris Tanenbaum of BTL in January 1954, whereas the first batch of commercially available silicon transistors were manufactured by Dr. Gordon Kidd Teal of Texas Instruments (TI) in the same year. In 1955, the first diffused silicon transistor made its appearance. To reduce the resistivity of the collector, the transistor with an epitaxial layer

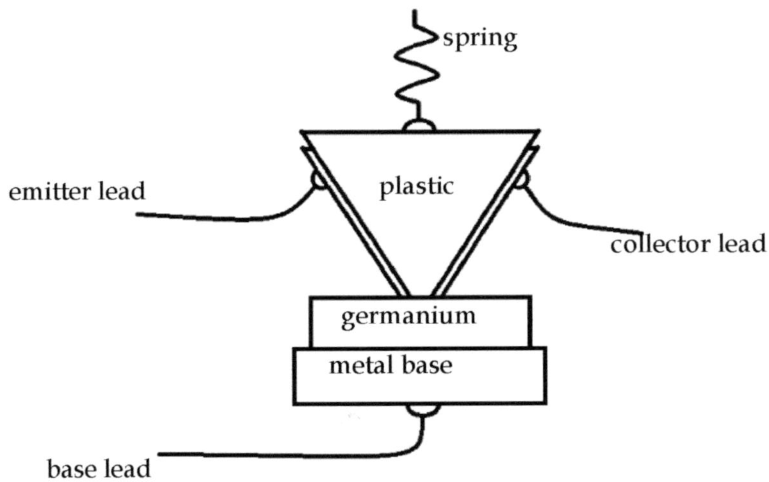

Figure 1.
An early model of the point-contact transistor.

deposited onto it was developed in 1960. In the same year, the planar transistor was proposed by Dr. Jean Amedee Hoerni [4, 7].

In 1958, Jack St. Clair Kilby, who was then an engineer in TI, successfully developed the first integrated circuit (IC). The device was just a simple 0.5 inch germanium bar, with a transistor, a capacitor and three resistors connected together using fine platinum wires. About half a year later in 1959, Dr. Robert Norton Noyce from Fairchild Camera (also one of the cofounders of Intel Corporation) invented independently his own IC chip. The interconnection in Noyce's 4 inch silicon wafer was realized by means of etching the aluminum film which was first deposited onto a layer of oxide [7]. Both Kilby and Noyce shared the patent right for the invention of the integrated circuit. In 2000, Kilby was awarded the Nobel Prize in Physics "for his part in the invention of the integrated circuit."

The first-generation computers were made of vacuum tubes. Conceived in 1937, the Atanasoff-Berry computer (ABC) (which was generally regarded as the first computer by many) and the Electronic Numerical Integrator and Computer (ENIAC) (which was credited as the first general purpose computer) built in late 1945 belong to the first-generation computers. The vacuum tube triode was swiftly replaced by the solid-state transistor since its advent in 1947. The second-generation computers were therefore made of transistors. The prototype computer built at the University of Manchester in November 1953 was widely regarded as the first transistor computer. Like the first transistor computer, most other electronic devices built before 1960 were actually based on germanium transistors. Although silicon transistors had already been developed in the mid-1950s, design engineers were more prone to using germanium than silicon. This is because the technology of germanium devices was very well established at that time and the reliability and yields of the silicon transistor were nowhere close to its germanium brethren [8]. Also, germanium switching diodes exhibited lower threshold voltage than silicon devices, allowing electronic devices made from germanium to be switched on at lower voltage drops [8]. The normal operating temperature of germanium devices, however, did not exceed 70°C, whereas silicon devices could operate well beyond 125°C. Hence, silicon was only used by the military establishment at that time for applications which were to operate at high temperature [8]. Besides its tenacity in withstanding temperature, silicon is also found to have lower leakage voltage and higher thermal conductivity than germanium [8]. These, however, were not the precipitating factors for silicon to replace germanium. It was the development of the oxide masking technique by Carl John Frosch and Lincoln Derick of BTL in 1955 which marked the pivoting point for the role played between silicon and germanium. Researchers found that an oxide film could be easily grown on the surface of a silicon substrate, but attempts to grow a stable oxide layer onto the germanium surface ended in total dismay [8]. BTL ceased meaningful research on germanium since then, and by the end of 1960, most of the electronic devices have been dominated by silicon. During these 5 years, researchers achieved several major technological innovations with the applications of the oxide films—some of which include the fabrication of the monolithic integrated circuit and the invention of the metal oxide semiconductor field-effect transistor (MOSFET) [8]. These innovations reinforce the status of silicon as the key element in electronic devices. In 1961, the first computer made from silicon IC chips was dedicated to the US Air Force.

Since the advent of the solid-state transistor and the demonstration on the workability of the IC chip about some 70 years ago, the electronic industries have been prospering hitherto. IC chips are now closely interwoven with human's life. They have, in many aspects, become indispensable to mankind. Indeed, one can easily find traces of IC chips intermingle into areas which intertwine seamlessly with the fabric of mankind's living hood. Some of these areas include transportation, telecommunication, security, medicine and entertainment, just to name a few.

3. Moore's law

In the article published in April 1965, one of the cofounders of Intel Corporation, Dr. Gordon Earle Moore, predicted that the number of electronic components (which include not just transistors but capacitors, resistors, inductors, diodes, etc. as well) in an IC chip would double every year [9]. Ten years later, Moore revised his prediction to a doubling of every 2 years. Moore's prediction, which is more commonly known as Moore's law nowadays, has been widely used by the IC manufacturers as a tool to predict the increase of components in a chip for the coming generations [10]. To date, Moore's law has been proven to have held valid for close to half a century. **Table 1** tabulates the progressive trend of the integration level for the semiconductor industry. It can be observed from the table that the number of transistors that can be fabricated in a chip has been growing continuously over the years. In fact, this growth has been in close agreement with Moore's law. In order to highlight the technological advancement in the IC industries, each decade since the inception of the semiconductor transistor has been earmarked as a different era. Eight eras have existed hitherto—they are the small-scale integration (SSI), medium-scale integration (MSI), large-scale integration (LSI), very large-scale integration (VLSI), ultra-large-scale integration (ULSI), super large-scale integration (SLSI), extra-large-scale integration (ELSI) and giant large-scale integration (GLSI) eras. During the VLSI era, a microprocessor was fabricated for the first time into a single IC chip. Although this era has now long passed, the VLSI term is still being commonly coined today. This is partly due to the absence of a significant qualitative leap between VLSI and its subsequent eras, and partly, it is also because IC engineers have been so used to this term; they decided to continue adopting it.

Integration level	Year	Number of transistors in a chip
Small-scale integration (SSI)	Late 1940s to late 1950s	Less than 100
Medium-scale integration (MSI)	Late 1950s to late 1960s	Between 100 and 1000
Large-scale integration (LSI)	Late 1960s to late 1970s	Between 1000 and 10,000
Very large-scale integration (VLSI)	Late 1970s to late 1980s	Between 10,000 and 100,000
Ultra-large-scale integration (ULSI)	Late 1980s to late 1990s	Between 100,000 and 1000,000
Super large-scale integration (SLSI)	Late 1990s to late 2000s	Between 1000,000 and 10,000,000
Extra-large-scale integration (ELSI)	Late 2000s to late 2010s	Between 10,000,000 and 100,000,000
Giant large-scale integration (GLSI)	Late 2010s to late 2020s	More than 100,000,000

Table 1.
Integration level of an integrated circuit chip.

4. The field-effect transistors

Today, the transistors fabricated in an IC chip are mostly MOSFETs. The earliest paper describing the operation principle of a MOSFET can be traced back to that reported in Julius Edgar Lilienfeld's patent in 1933 [11]. Unfortunately, the technology at that time was inadequate to allow Lilienfeld's idea to be physically materialized. In 1959, Dr. Dawon Kahng and Dr. Martin M. (John) Atalla at the BTL successfully constructed the MOSFET [12]. In 1963, two engineers from the Radio Corporation of America (RCA) Princeton laboratory, Dr. Steven R. Hofstein and Dr. Frederic P. Heiman, presented the theoretical description on the fundamental nature of the silicon planar MOSFET [13]. In the same year, Dr. Tom Chih-Tang Sah and Dr. Frank Marion Wanlass of Fairchild Semiconductor invented the first complementary metal oxide semiconductor (CMOS) logic circuit [14]. In 1989, Dr. Digh Hisamoto and his team member at Hitachi Central Research Laboratory introduced the fin field-effect transistor or better known as the FinFET—a non-planar MOSFET modified from its planar counterpart. Although the FinFET was found to possess various advantages over the planar MOSFET, it was not adopted by the industries then. This was partly due to the difficulty in fabricating its three-dimensional structure and, partly, also because the planar MOSFETs still had plenty of rooms to be improved further. Having realized that the planar MOSFET was gradually approaching its bottleneck in its technological advancement, chipmakers started to resort to FinFETs in the fabrication of high-end electronic devices (such as microprocessors) in 2011.

4.1 The MOSFET

The MOSFET is nothing more than a device which operates as an electronic switch. **Figure 2** shows the basic structure of the MOSFET. The transistor comprises four terminals, namely, the drain (D), source (S), gate (G) and substrate or body (B) terminals. As can be clearly seen from the figure, the device constitutes three layers—a polysilicon layer (which forms the gate terminal), an oxide layer (known the gate oxide) and a single-crystal semiconductor layer (known as the substrate). In the early days, the gate terminal was made of aluminum. It is from these three layers of materials that the FET device acquired its name. In the mid-1970s, however, the gate material was replaced with polysilicon. When ion implantation was introduced to form the self-aligned source and drain terminals in the 1970s, a high-temperature (higher than 1000°C) annealing process was required to repair the damaged crystal structure at the surface of the substrate, as a result of the energetic dopant ion bombardment and to activate the dopant [15]. IC engineers observed that the aluminum gate melted during the annealing process. This is because aluminum has a melting point of about 660.3°C. In order to overcome this problem, polysilicon which has a melting point of about 1414°C was employed as the replacement for gate material. Although the gate today is no longer made of aluminum, the term MOSFET has been so widely accepted that it stays until today.

The basic operation principle of a MOSFET is actually quite straightforward. When a voltage source is connected in between the drain and source terminals, a conducting channel is to be formed between the two terminals to allow the current to flow. The channel is commonly referred to as the inversion layer since the charges accumulated at the channel oppose those of the substrate. In this case, the gate terminal acts like a switch which controls the formation of the inversion layer. When sufficient voltage drop (and, of course, with the appropriate polarity) is applied to the gate terminal, carriers would be attracted to the gate oxide-substrate interface to form the inversion layer.

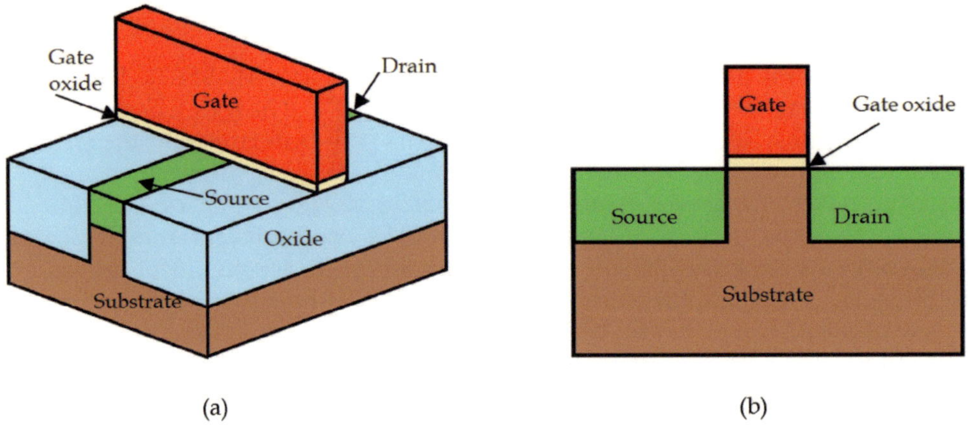

Figure 2.
The (a) basic structure and (b) cross section of a MOSFET.

A MOSFET can be classified into two types, depending on the dopants in the drain and source terminals, as well as the substrate. When both the drain and source terminals, in a p-type substrate, are heavily doped with donator ions (such as phosphorous or arsenic), a negative channel is to be formed in between them to conduct current. On the other hand, when both terminals, in an n-type substrate, are heavily doped with acceptor ions (such as boron), a positive channel is to be formed. The former device is therefore known as a negative channel MOSFET or an NMOS transistor, while the latter is known as a positive channel MOSFET or a PMOS transistor. **Figure 3** shows the circuit symbols of both PMOS and NMOS transistors [4].

The size of a MOSFET transistor is measured by the gate length, which is also commonly known as the feature size or feature length as is denoted by the symbol L. The size of the transistor has been shrinking tremendously over the years. This allows a higher number of transistors to be fitted into a single die. Overseen by the Taiwan Semiconductor Industry Association (TSIA), the US Semiconductor Association (SIA), the European Semiconductor Industry Association (ESIA), the Japan Electronics and Information Technology Industries Association (JEITA) and the Korean Semiconductor Industry Association (KSIA), the International Technology Roadmap of Semiconductor (ITRS) is charted to forecast how the technology node is expected to evolve. The purpose of the ITRS is to ensure healthy growth of the IC industries. **Table 2** tabulates the progressive reduction of the feature size published in ITRS 2.0 [16]. In order to provide a clear outline to simplify academic, manufacturing, supply and research coordination regarding the development of electronic devices and systems, the ITRS was continued by the International Roadmap for Devices and Systems (IRDS) in 2018 [17].

4.2 The FinFET

As the feature size reduces to the submicron regimes, fields at the source and drain regions become comparatively high, and this may induce certain adverse effects to the charge distribution. Some of the examples of these short-channel effects are the threshold voltage roll-off in the linear region, drain-induced barrier lowering (DIBL) and bulk punch-through [18]. To suppress these effects, additional steps such as the introduction of retrograde well, the deposition of the sidewall spacers, lightly doped drain (LDD) implantation, halo implantation, etc. have been introduced into the IC fabrication process [19]. As the device continues to shrink,

Figure 3.
The symbol of (a) a PMOS transistor and (b) an NMOS transistor.

Physical gate length	Year						
	2015	2017	2019	2021	2024	2027	2030
High-performance logic (nm)	24	18	14	10	10	10	10
Low-performance logic (nm)	24	20	16	12	12	12	12

Table 2.
Forecast of gate length by ITRS.

curbing the short-channel effects turns out to be a strenuous task. When the feature size approaches the subnanometer range (i.e. 90 nm and below), static leakage current due to the short-channel effects has become a serious problem.

When the technology node reached 22 nm in 2011, Intel Corporation announced the fabrication of the tri-gate transistor, replacing the conventional planar MOSFET. Better known as the FinFET, this device has a three-dimensional transistor structure, as depicted in **Figure 4** [20]. It is apparent from the figure, a FinFET is named so because of the protruding source and drain terminals from its substrate surface, which resemble the fins of a fish. Since the gate wraps around the inversion layer, FinFETs provide higher current flow from the source to the drain terminals. This protruding fin structure also allows better control of the current flow, i.e. it reduces current leakage considerably when the device is at its "off-state" and minimizes short-channel effects at its "on-state". Since the device has lower threshold voltage than the planar MOSFET, a FinFET can also operate at relatively lower voltage drops. In a nutshell, the FinFET shows less leakage, faster switching and lower power consumption in comparison to its planar counterpart.

5. IC design flow

Generally, the design process of an IC chip involves three stages—namely, the (i) behavioral, (ii) logic circuit and (iii) layout representations [4, 21]. At the end of each stage, verification is to be performed before proceeding to the next. Hence, it is common to have repetitions and iterations in the processes [4, 21].

5.1 Behavioral representation

At the initial stage of IC design, it is important to be specific on the functionalities of the chip. The design architecture is to be drawn out. Verilog or SystemVerilog hardware description language (HDL) is used to define the behavior of the IC chip.

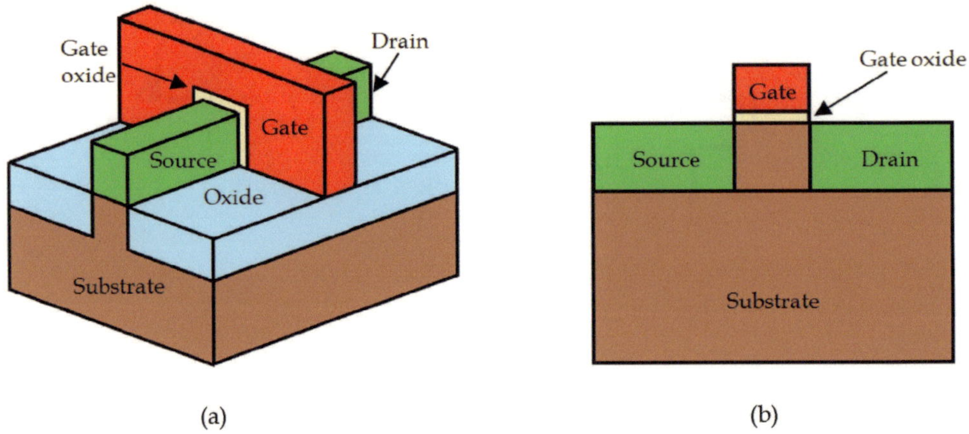

Figure 4.
The (a) basic structure and (b) cross section of a FinFET.

5.2 Logic circuit representation

Once the HDL codes are successfully simulated, functional blocks from standard cell libraries are used to synthesize the behavioral representation of the design into logic circuit representation. Once the design is verified, the gate-level netlist is generated. The netlist consists of important information of the circuit such as the connectivity and nodes and is necessary in order to develop the layout of the design.

5.3 Layout representation

The physical layout of the design is created at the final stage. The process starts with floor planning which defines the core and routing areas of the chip. In order to optimize the design, the building blocks are usually adjusted and orientated by IC designers. This process is known as placement. Once this is completed, a routing process is performed to interconnect the building blocks.

6. Microchip fabrication

To fabricate the chip, the layout is sent to a fab or a foundry. In a fab, a single-crystal semiconductor ingot is first grown. Wafers are then sliced from the ingot. The layout is printed onto the dice in each wafer.

The fabrication process for NMOS and PMOS transistors is similar. The main differences lie within the types and density of dopants applied to the substrate—specifically in the formation of well, threshold voltage V_{TH} adjust implantation, LDD implantation, source/drain implantation, etc. The process flow of fabricating a planar MOSFET is summarized in the following sections, and it is also graphically depicted in **Figure 5**. The process of chip fabrication can be broadly separated into five stages, i.e. (i) well formation, (ii) device isolation, (iii) transistor making, (iv) interconnection and (v) passivation [15].

6.1 Well formation

Initially, a p-type single-crystal silicon wafer is prepared (**Figure 5(i)**). In order to form a P (for NMOS) or N (for PMOS) well, screen oxide is first grown on the surface of the substrate (**Figure 5(ii)**). A high-energy ion implantation is then

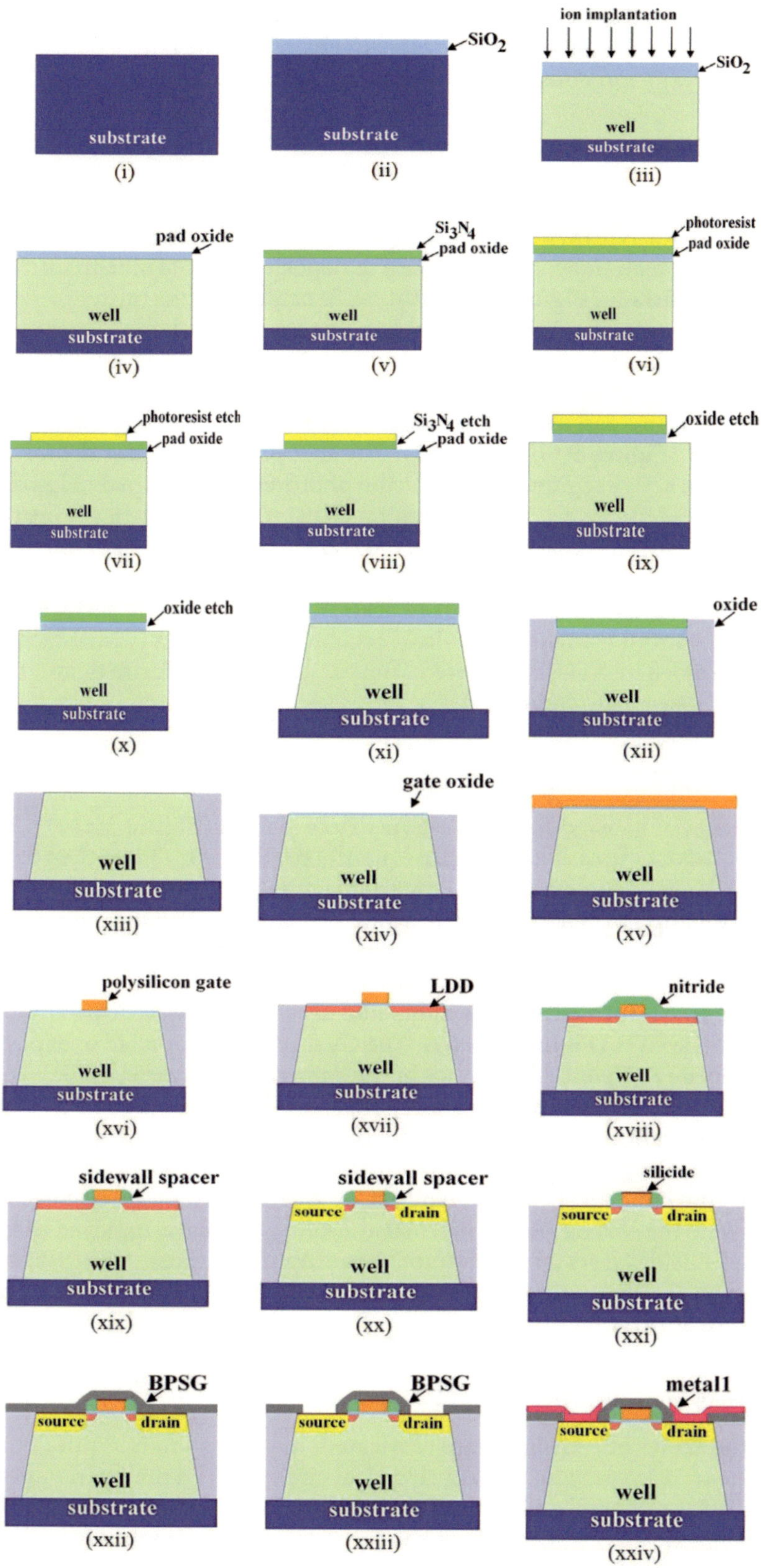

Figure 5.
The fabrication process of a MOSFET.

performed to form the well (**Figure 5(iii)**). The wafer subsequently undergoes annealing and drive-in processes to, respectively, repair the lattice damage caused by the high-energy ion bombardment and to activate the dopant.

6.2 Device isolation

Next, shallow trench isolation STI is employed to isolate neighboring devices. Initially, pad oxide is grown via dry oxidation (**Figure 5(iv)**). Chemical vapor deposition CVD technique is then applied to deposit a layer of silicon nitride Si_3N_4 onto the oxide surface (**Figure 5(v)**). Pad oxide acts as a stress buffer to avoid cracks on the nitride film, whereas nitride film acts as a mask for silicon etching. A layer of photoresist is subsequently deposited onto the nitride layer (**Figure 5(vi)**). Lithography is performed to develop patterns on the photoresist (**Figure 5(vii)**). The nitride film and pad oxide are etched in accordance with the pattern formed at the photoresist (**Figure 5(viii)**and **(ix)**). The area protected under the nitride mask is known as the active region. As soon as the photoresist is stripped (**Figure 5(x)**), the substrate undergoes reactive ion etching (RIE) to form trenches (**Figure 5(xi)**). A thin layer of barrier oxide is grown in the trenches so as to block impurities from diffusing into the substrate during the CVD process. The trenches are then filled with oxide via the CVD process (**Figure 5(xii)**). The oxide at the surface of the substrate is removed using the chemical mechanical polishing (CMP) technique (**Figure 5(xiii)**). The STI is completed after annealing is performed, and the nitride and pad oxide layers are etched.

6.3 Transistor making

A thin layer of gate oxide is applied via dry oxidation (**Figure 5(xiv)**). Threshold voltage V_{TH} adjust implantation is subsequently performed. This is then followed by thermal annealing to repair the lattice damage at the substrate surface. A layer of polysilicon is deposited onto the substrate surface after the annealing process (**Figure 5(xv)**). The polysilicon is then etched according to the dimension of the feature size and annealed to form the polysilicon gate.

Once the gate is formed, LDD is implanted to suppress hot electron effect in deep submicron MOSFETs (**Figure 5(xvii)**). The CVD process is applied to deposit a layer of silicon nitride Si_3N_4 onto the surface of the substrate (**Figure 5 (xviii)**). The nitride film is etched to form sidewall spacers at both sides of the gate (**Figure 5 (xix)**). The source/drain dopant is then implanted into the substrate. The substrate subsequently undergoes annealing after the implantation process (**Figure 5(xx)**). This is then followed by the removal of the thin oxide layer. A layer of titanium or cobalt is then deposited onto the surface. Rapid thermal annealing (RTA) is employed to form the self-aligned silicide layers on the gate and source/drain surfaces. At the final stage of the transistor fabrication process, the unreacted titanium or cobalt layer is etched away (**Figure 5(xxi)**).

6.4 Interconnection

Once the arrays of transistors are fabricated, metallization is required to interconnect the transistors so as to form electrical circuitries. In the interconnection stage, a layer of premetal dielectric (PMD) is first formed by depositing a layer of borophosphosilicate glass BPSG onto the substrate surface (**Figure 5(xxii)**). The PMD acts as the first layer of insulator for multilevel interconnection. After the die is annealed, the BPSG is etched to form source/drain contacts (**Figure 5(xxiii)**). Metallization is applied by depositing and etching aluminum (Al) on the contacts

(**Figure 5(xxiv)**). Phosphosilicate glass PSG is used as the insulator material for the subsequent levels of metal interconnections. The insulator layers after PMD is known as the intermetal dielectric (IMD) layers. Vials filled with tungsten are usually used to interconnect different levels of metal layers.

6.5 Passivation

The passivation layer is the final dielectric layer deposited onto the die after the last metal interconnection is formed. Silicon nitride is usually used as the passivation layer.

7. Packaging

To protect the chip from harsh external environment (e.g. being exposed to UV light or moisture or being scratched), it is essential to encapsulate the chip in a ceramic or plastic package—a process known as packaging. The three most commonly used packaging techniques are (i) wire bonding, (ii) flip chip and (iii) tape-automated bonding (TAB) [10]. IC packaging marks the end of the entire chip manufacturing process. The chip is therefore ready to be released to the market, once the packaging process is completed.

Author details

Kim Ho Yeap[1*], Muammar Mohamad Isa[2] and Siu Hong Loh[1]

1 Universiti Tunku Abdul Rahman, Jalan Universiti, Kampar, Perak, Malaysia

2 Universiti Malaysia Perlis, Jalan Wang Ulu Arau, Kangar, Perlis, Malaysia

*Address all correspondence to: yeapkh@utar.edu.my

References

[1] Nordrum A. Another step toward the end of Moore's law. IEEE Spectrum. 2019;**6**:9-11

[2] Moore SK. TSMC's 5-nanometer process on track for first half of 2020. BACUS. 2020;**36**(1):8

[3] Hiramoto T. Five nanometre CMOS technology. Nature Electronics. 2019;**2**:557-558. DOI: 10.1038/s41928-019-0343-x

[4] Yeap KH, Nisar H. Introductory Chapter: VLSI. In: Yeap KH, Nisar H, editors. Very-Large-Scale Integration. Rijeka, Croatia: InTechOpen; 2018. pp. 3-11

[5] Yeap KH, Introductory Chapter NH. Complementary metal oxide semiconductor (CMOS). In: Yeap KH, Nisar H, editors. Complementary Metal Oxide Semiconductor. London, UK: IntechOpen; 2018. pp. p3-p7

[6] Arsov GL. Celebrating 65th anniversary of the transistor. Electronics. 2013;**17**:63-70. DOI: 10.7251/ELS1317063A

[7] Lukasiak L, Jakubowski A. History of semiconductors. Journal of Telecommunications and Information Technology. 2010;**1**:3-9. DOI: 10.1088/0031-9120/40/5/002

[8] Seidenberg P. From germanium to silicon: A history of change in the technology of the semiconductors. In: Goldstein A, Aspray W, editors. New Brunswick: IEEE Press; 1997. pp. 35-74

[9] Moore GE. Cramming more components onto integrated circuits. Electronics. 1965;**38**:14-117. DOI: 10.1109/N-SSC.2006.4785860

[10] Yeap KH. Fundamentals of Digital Integrated Circuit Design. 1st ed. UK: Authorhouse; 2011

[11] Lilienfeld JA. Method and apparatus for controlling electric currents. U. S. Patent No. 1745175A (Filed: 08 October 1926. Issued: 28 January 1930)

[12] Kahng D. Electric field controlled semiconductor device. U. S. Patent No. 3, 102, 230 (Filed: 31 May 1960. Issued: 27 August 1963)

[13] Hofstein SR, Heiman FP. Silicon insulated-gate field-effect-transistor. Proceedings of the IEEE. 1963;**51**:1190-1202

[14] Wanlass SM, Sah CT. Nanowatt logic using field-effect metal-oxide semiconductor triodes. In: Proceedings of the IEEE Conference on Solid-State Circuits Conference. Digest of Technical Papers; US; 1963. pp.32-33

[15] Xiao H. Introduction to Semiconductor Manufacturing Technology. US: Prentice Hall; 2001

[16] International Technology Roadmap for Semiconductors 2.0 [Internet]. 2015. Available from: https://www.semiconductors.org [Accessed: 03 April 2017]

[17] International Roadmap for Devices and Systems [Internet]. Available from: https://irds.ieee.org/

[18] Sze SM. Semiconductor Devices: Physics and Technology. 2nd ed. US: John Wiley and Sons; 2002

[19] Ahmad I, Ho YK, Majlis BY. Fabrication and characterization of a 0.14 μm CMOS device using ATHENA and ATLAS simulators. International Scientific Journal of Semiconductor, Physics, Quantum Electronics, and Optoelectronics. 2006;**9**(2):40-44. DOI: 10.15407/spqeo

[20] Yeap KH, Lee JY, Yeo WL, Nisar H, Loh SH. Design and characterization

of a 10 nm FinFET. Malaysian Journal of Fundamental and Applied Sciences. 2019;**15**(4):609-612

[21] Yeap KH, Thee KW, Lai KC, Nisar H, Krishnan KC. VLSI circuit optimization for 8051 MCU. International Journal of Technology. 2018;**9**(1):142-149

Section 2

Microchip Design Methods

Chapter 2

Ultra-Low-Voltage IC Design Methods

Daniel Arbet, Lukas Nagy and Viera Stopjakova

Abstract

The emerging nanoscale technologies inherently offer transistors working with low voltage levels and are optimized for low-power operation. However, these technologies lack quality electronic components vital for reliable analog and/or mixed-signal design (e.g., resistor, capacitor, etc.) as they are predominantly used in high-performance digital designs. Moreover, the voltage headroom, ESD properties, the maximum current densities, parasitic effects, process fluctuations, aging effects, and many other parameters are superior in verified-by-time CMOS processes using planar transistors. This is the main reason, why low-voltage, low-power high-performance analog and mixed-signal circuits are still being designed in mature process nodes. In the proposed chapter, we bring an overview of main challenges and design techniques effectively applicable for ultra-low-voltage and low-power analog integrated circuits in nanoscale technologies. New design challenges and limitations linked with a low value of the supply voltage, the process fluctuation, device mismatch, and other effects are discussed. In the later part of the chapter, conventional and unconventional design techniques (bulk-driven approach, floating-gate, dynamic threshold, etc.) to design analog integrated circuits towards ultra-low-voltage systems and applications are described. Examples of ultra-low-voltage analog ICs blocks (an operational amplifier, a voltage comparator, a charge pump, etc.) designed in a standard CMOS technology using the unconventional design approach are presented.

Keywords: analog/mixed-signal IC design, unconventional design approach, bulk-driven design, ultra-low-voltage, ultra-low-power, standard nanoscale CMOS technology

1. Introduction

The design of ultra-low-voltage (ULV) and low-power (LP) analog and mixed-signal ICs in modern nanotechnologies represents a real challenge for circuit designers and researches, since it introduces several limitations in numerous aspects. Firstly, since advanced nanoscale technologies offer a possibility to design analog, digital, and radio-frequency (RF) circuits as well as micro-electro-mechanical systems (MEMS) on a single chip, there is usually issue of a common value of the supply voltage. With the technology development, the value of the supply voltage is scaled down significantly. However, the threshold voltage (V_{TH}) of the MOS devices is *not* lowered at the same pace. This fact reduced the voltage headroom for conventional circuit topologies (e.g., cascode structures) to operate correctly. Low value of the supply voltage may significantly influence the main parameters of analog ICs such as dynamic range (DR), power supply rejection (PSR), noise immunity, etc. The second limiting factor lies in the significant fluctuation of process parameters in nanoscale technologies that brings new requirements to IC design—circuits have to be robust enough against process, temperature, and voltage variations [1].

From the IC design point of view, one of the main problems caused by a lowered V_{DD} value is the reduction of useful voltage range for existing and standard circuit topologies. Analog circuits are suffering mostly from this limiting drawback. Decreasing the threshold voltage, as well as thinner layer of the gate oxide of a MOS (metal oxide semiconductor) transistor cause steep rising of the sub-threshold leakage current that is rather typical for nanotechnologies. These reasons do limit the further decrease of the threshold voltage. **Figure 1** depicts the dependency of the V_{DD} level and the threshold voltage on the technology node that is predicted for years to come by IRDS (International Roadmap for Devices and Systems). One can observe that the threshold voltage cannot follow the trend of the supply voltage level decrease due to substantial leakage currents.

The minimum power supply voltage of CMOS analog ICs designed without dedicated low-voltage (LV) techniques is limited by a value given by the sum of the turn-on voltage V_{GS} of MOS transistor and required voltage swing. For example, the voltage of ≈ 300 mV can be considered an average threshold voltage level in standard deep sub-micron CMOS fabrication process for transistors with reasonable channel length. This amount of external voltage applied between the gate and bulk terminal (or vice versa) is usually sufficient to introduce a strong inversion in the MOS structure and hence, turn-on the transistor. Another problem created by low supply voltages ($V_{DD} \approx$ 600 mV and lower) is the limited voltage headroom for cascode circuit structures and stacked transistors [2]. Therefore, new design approaches focused on the low-voltage circuit topologies that can overcome limitations mentioned above are still required.

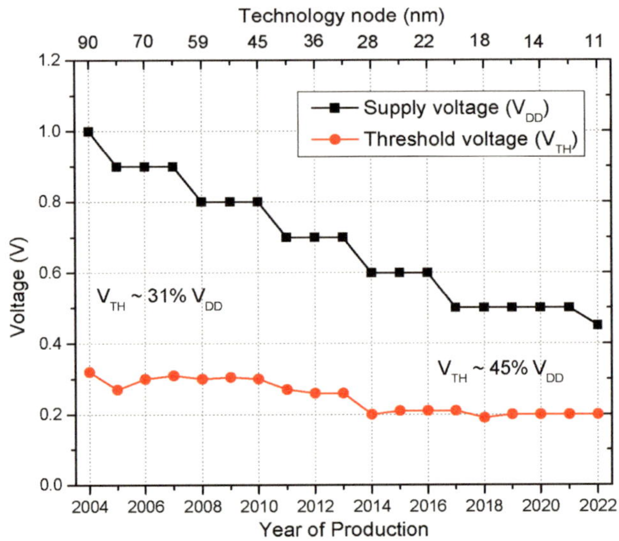

Figure 1.
Scaling the supply voltage and threshold voltage in time.

2. Low-voltage design techniques and approaches

In this section, the survey of low-voltage design techniques and approaches that can be used in a standard CMOS technology (no additional process steps) are presented. Generally, low-voltage design techniques can be divided into two groups: conventional methods and unconventional ones. Unconventional methods include bulk-driven (BD) approach, dynamic threshold technique, floating-gate method, quasi-floating gate, and bulk-driven quasi-floating gate approaches. However, only the circuits designed by the bulk-driven and dynamic threshold approaches can be implemented in the standard CMOS technologies without any modification of the fabrication process. On the other hand, the conventional techniques such as circuits with rail-to-rail input/output operating range, MOS transistors working in sub-threshold region, level shifter techniques or MOS transistor in self-cascode structure represent commonly used approaches in the area of low-voltage IC design.

Since only circuits designed by the bulk-driven approach can be implemented in pure CMOS technology, in this chapter, we focus on this LV circuit design technique. At the end of this chapter, some examples of experimental and silicon-proven analog/mixed-signal circuits designed by the BD approach are presented.

2.1 MOS transistor in sub-threshold operation region

Firstly, it is vital to explain the operation regions of the MOS transistor, since this is the most important aspect for analog IC design. The optimum IC design is characterized by the minimum power consumption, minimum silicon area and sufficient frequency response, gain and other circuit specifications. Analog and mixed-signal circuit design procedure of systems using (ultra) low-power supply voltage introduces an extra layer of challenges for even seasoned circuit designers. The problems low supply voltage introduces, negatively influence several design considerations, circuit attributes and possible design options. The first and foremost is the substantially limited inversion level the MOS transistors operate in. This results, among others, in higher mismatch between transistor parameters, exponential temperature sensitivity, and drastically lowered operational frequency. We must not forget the increased silicon area requirements due to large transistors compensating for low transconductance values, increased noise and difficulties with precise secondary effects modeling. All of the above are typical drawbacks of low-voltage/low-power circuit design and their application [3]. The second issue is topological. It lies in constrained possible number of stacked transistors, in order to ensure their operation in saturation region. According to [4], the *theoretical* lower limit for saturation voltage of a MOS transistor in deep the sub-threshold region is defined as $V_{DSsat(min)} \approx 4 \cdot \frac{kT}{q}$, which at room temperature, equals to approximately 105 mV. However, with increasing inversion level, this value grows with square root trend.

The situation has been greatly improved by the development of design-oriented charge-sheet based EKV MOS transistor model (named after its authors—Enz-Krummenacher-Vittoz) [5]. EKV model defines the parameters of MOS device dependent on continuous range of inversion level unlike the industry-standard threshold voltage-based BSIM models. EKV model also introduced the so-called g_m/I_D design approach, which avails simple, yet accurate hand-calculations, straightforward transistor sizing and complete technology independence. In [4], the author defines the level of inversion, also called *inversion coefficient* (IC) of a MOS structure by Eq. (1).

$$IC = \left[ln\left(1 + e^{\frac{V_{GS}-V_{TH}}{2.n.U_T}}\right)\right]^2 = \frac{I_D}{I_{technology} \cdot \frac{W}{L}}, \tag{1}$$

where V_{GS} is voltage between gate and source terminal of the MOS transistor, V_{TH} is a MOS transistor's threshold voltage, n is sub-threshold so-called slope factor, U_T is Boltzmann's thermal voltage (25.86 mV at room temperature), W is MOS transistor channel width, L is MOS transistor channel length, and $I_{specific}$ technology specific, current when a square MOS device (W = L) is in the middle of inversion range IC = 1.

The point when IC = 1 also determines the conditions when the drain diffusion current equals drain drift current. The interpolated dependency of *transconductance efficiency*—g_m/I_D as a function of IC is defined by Eq. (2). It represents very powerful formula since it is completely technology independent [6]. Furthermore, it can be easily implemented into a spreadsheet, introducing automated calculations and transistor sizing.

$$\frac{g_m}{I_D} = \frac{1}{n.U_T.\left(\frac{1}{2} + \sqrt{\frac{1}{4} + IC}\right)} \tag{2}$$

Figure 2 depicts the dependency of g_m/I_D on the inversion coefficient—IC, governed by Eq. (2). The area where IC $\leq$ 0.1 represents the sub-threshold operation region that is also called *weak inversion*. The MOS transistor operating under these conditions exhibits high voltage gain, low drain current, low saturation voltage but also large dimensions in order to compensate low transconductance and very low cut-off frequency. When inversion coefficient becomes IC $\geq$ 10, the MOS devices is operating in *strong inversion* or above the threshold voltage—the traditional working conditions. MOS transistor can process signals at high frequencies and does not require much of silicon area. However, the gain lowers and the drain current increases. The region in between of the weak and strong inversion (0.1 $\leq$ IC $\geq$ 10) describes a smooth transition between these two states. It is often called a *moderate inversion* and it represents a very good trade-off of the transistor and circuit parameters. Furthermore, the modern nanoscale CMOS technologies, working with lowered power supply voltage, are shifting the transistor operation into the

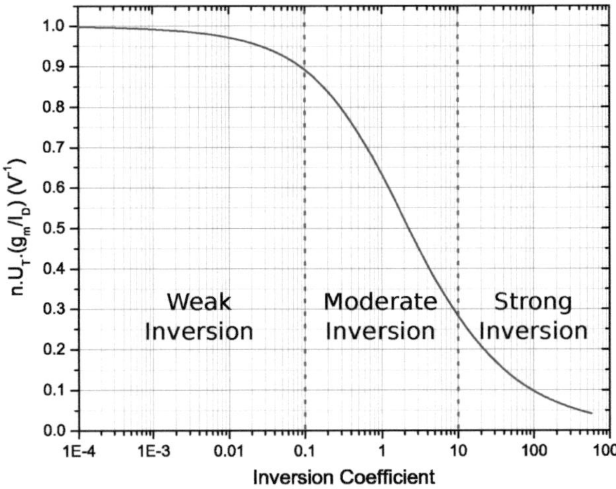

Figure 2.
g_m/I_D as a function of inversion coefficient—IC.

moderate inversion, as the voltage headroom decreases and the level of the threshold voltage remains fairly constant over time (**Figure 1**).

2.2 Bulk-driven design approach

In the conventional approaches, MOS transistor is usually controlled by its gate potential. However, the current flowing through the device can also be modulated by the bulk-source voltage V_{BS}, which is usually considered a parasitic effect and may introduce undesired body transconductance g_{mb}. In the BD design approach, the input signal is applied to the transistor bulk, while a bias voltage is connected to the gate in order establish a channel between the source and drain terminals. If a constant V_{GS} is kept as the bias voltage and the input signal is applied to the bulk electrode, then a JFET-like transistor behavior can be obtained. In other words, the inversion channel width is modulated according to the voltage applied to the bulk. Using the bulk as the signal input results in significantly reduced need to overcome the threshold voltage at the MOS transistor.

The effect of the V_{BS} on the drain current is embedded in the threshold voltage V_{TH}. The threshold voltage of MOS transistor can be expressed by Eq. (3). It also serves as *very important* link between g_m/I_D, IC and bulk-driven design approaches.

$$V_{TH} = V_{TH0} \pm \gamma\left(\sqrt{2|\Phi_F| - V_{BS}} - \sqrt{2|\Phi_F|}\right), \tag{3}$$

where V_{TH0} is the threshold voltage with V_{BS} = 0 V, γ is technology-specific body factor, and Φ_F is technology-specific Fermi's potential.

Thus, changes in V_{BS} value will result in modification of V_{TH}, which will inevitably modify the inversion coefficient according to Eq. (1) and finally, control the MOS transistor drain current.

In order to analyze the properties of a MOS transistor driven by the bulk terminal, the conventional gate-driven and bulk-driven single stage common-source amplifiers, depicted in **Figures 3** and **4**, have been investigated and compared.

From **Figures 3** and **4**, it can be observed that the input capacitance of the BD single stage amplifier will be higher than in the case of the GD amplifier. It is caused by a parasitic capacitance C_{bsub} between the bulk and substrate terminals. In the case of the GD amplifier, the input capacitance depends on C_{gs} and C_{gd} capacitances, while the BD amplifier has the input capacitance dependent on the bulk-source C_{bs}, bulk-drain C_{bd} and bulk-substrate C_{bsub} capacitances combined together.

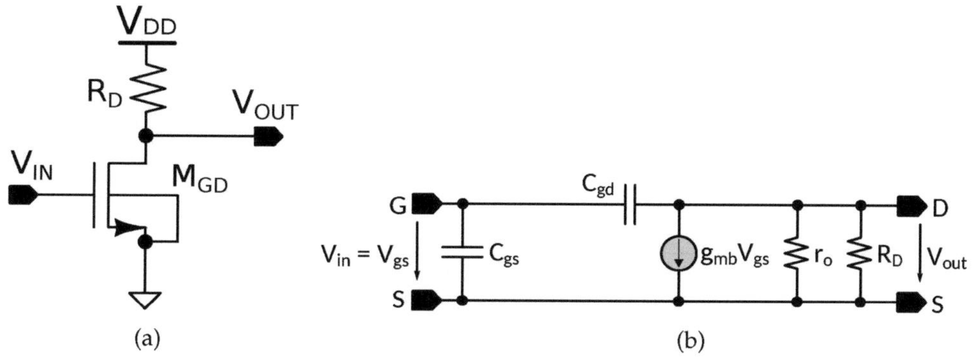

Figure 3.
Gate-driven common-source amplifier. (a) Schematic diagram, and (b) Small-signal model.

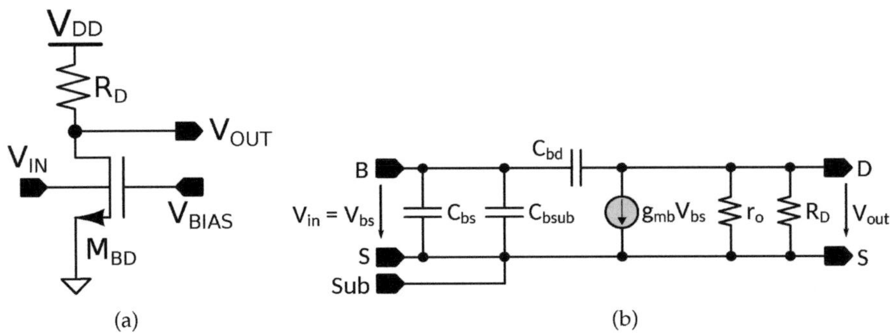

Figure 4.
Bulk-driven single stage amplifier. (a) Schematic diagram, and (b) Small-signal model.

The transconductance of the conventional GD transistor can be expressed by the following Eq. (4)

$$g_m = \beta \frac{W}{L} \cdot (V_{GS} - V_{TH}) \tag{4}$$

It is important to point out that Eq. (4) is only valid when the MOS transistor operates in the strong inversion. In the weak inversion, the transconductance is proportionally dependent on the drain current, as given by Eq. (5).

$$g_{m_{wi}} = \frac{I_{DS_{wi}}}{n \cdot U_T} \tag{5}$$

The relationship between the transconductance of a GD transistor g_m and BD transistor g_{mb} is given by Eqs. (4) and (6).

$$\begin{aligned} g_{mb} &= \frac{\gamma}{2\sqrt{|2\phi_F - V_{bs}|}} \cdot g_m \\ g_{mb} &= \frac{C_{btot}}{C_{gtot}} \cdot g_m \\ g_{mb} &\approx (0.2 \div 0.3) \cdot g_m \end{aligned} \tag{6}$$

where C_{btot} and C_{gtot} are the total parasitic capacitances between bulk and channel and gate and channel, respectively. It can be observed that transconductance of the BD MOS transistor is only 20–30% of transconductance of the GD MOS transistor.

In order to determine the frequency performance of the BD transistor, schematic diagram, and small-signal model (depicted in **Figure 5**) have to be employed. Using small-signal model, the transition frequency $f_{T,BD}$ of the BD MOS transistor can be obtained. Firstly, it is important to define the transition frequency of the GD MOS transistor given by Eq. (7).

$$f_{T,GD} = \frac{g_m}{2\pi C_{gs}} \tag{7}$$

The transfer function and current gain of the BD MOS transistor can be expressed by Eq. (8).

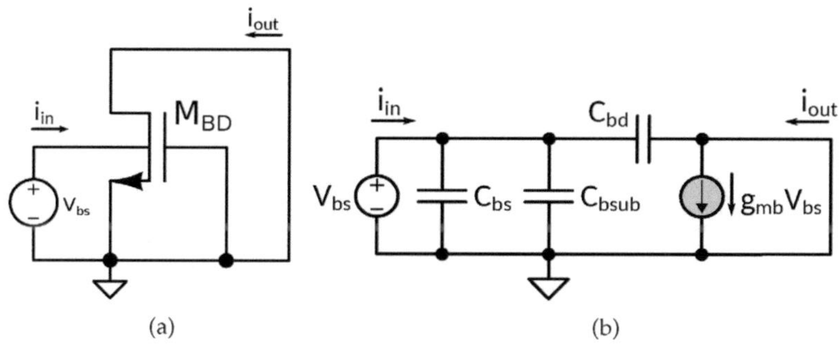

Figure 5.
Schematic diagram and small-signal model for $f_{T,BD}$ calculation. (a) Schematic diagram, and (b) Small-signal model.

$$\begin{aligned}\frac{i_{out}}{i_{in}} &\approx \frac{g_{mb}.V_{bs}}{j\omega(C_{bs}+C_{bsub}+C_{bd}).V_{bs}}\\ \frac{i_{out}}{i_{in}} &\approx \frac{g_{mb}}{j\omega(C_{bs}+C_{bsub}+C_{bd})}\end{aligned} \tag{8}$$

If we consider that unity small-signal gain is obtained at frequency $\omega_{T,BD}$, the transition frequency of the BD MOS transistor can be expressed as follows:

$$\begin{aligned}f_{T,BD} &= \frac{1}{2\pi}\cdot\omega_{T,BD}\\ f_{T,BD} &= \frac{g_{mb}}{2\pi(C_{bs}+C_{bsub}+C_{bd})}\\ f_{T,BD} &\approx (0.2\div 0.3)\cdot f_{T,GD}\end{aligned} \tag{9}$$

As can be observed from Eq. (9), the transition frequency of the BD MOS transistor is about five times lower than in the case of a MOS transistor driven by gate terminal. Another important parameter of the amplifier is the noise introduced into the circuit by the active component. The input referred noise of the GD MOS transistor depends on the current i_{ds} and transconductance g_m, and can be expressed as follows:

$$v_{noise}^2 = \frac{i_{ds}^2}{g_m^2}o \tag{10}$$

Similarly, the input referred noise of the BD MOS transistor is given by Eq. (11), where one can observe that the BD MOS transistor suffers from higher noise due to the lower transconductance g_{mb}.

$$v_{noise,BD}^2 = \left(\frac{g_m}{g_{mb}}\right)^2\cdot v_{noise}^2 \tag{11}$$

The small-signal output resistance for both GD and BD transistors is identical, and given by Eq. (12).

$$r_o = \frac{1}{\lambda I_{DS}} = \frac{V_A}{I_{DS}}, \tag{12}$$

where V_A represents early voltage and I_{DS} is the current flowing through the MOS transistor. As mentioned above, the BD technique uses bulk terminal for the signal input, which results in significantly reduced need to overcome the threshold voltage at the MOS transistor input, as a whole. In summary, we can state the important advantages of the BD design technique, which include the following:

- BD MOS transistor depletion characteristics significantly reduce the need to overcome the threshold voltage V_{TH} at the transistor input and increases the voltage headroom for low-voltage applications.
- Suitable for rail-to-rail voltage range.
- Better linearity due to low, transconductance (g_{mb}).
- Possibility to operate with a low value of the power supply.
- Easy to implement in a standard CMOS technology (twin-well process, both MOS devices available).

Unfortunately, if compared to traditional GD design approach, the bulk-driven design method also exhibits the following disadvantages:

- Body transconductance g_{mb} of the BD MOS transistor is 4–5 times lower than the gate transconductance g_m, which leads to inferior frequency response and decreased gain-bandwidth product.
- Input capacitance of the BD MOS transistor is greater, if compared to the traditional GD device.
- Input noise of the BD MOS transistor is increased.
- BD MOS transistors fabricated in a standard CMOS process are prone to the catastrophic latch-up effect.

The last drawback, however, can be effectively mitigated by lowering the power supply voltage below the threshold voltage of a PN junction or by usage of an expensive silicon-on-insulator (SOI) fabrication process. This step would prevent the turn-on of the parasitic bipolar transistor in the substrate.

3. Design examples of low-voltage analog ICs

In this section, several design examples and circuit topologies of basic analog IC building blocks using bulk-driven approach are presented. The described blocks have been silicon-proven through fabrication in a standard CMOS nanotechnology and measurement evaluation of the chip prototypes.

3.1 BD current mirrors

One of the most widely used circuit structures employed in IC design are arguably the current mirrors (CM). It is a two-port circuit, which processes the input current I_{REF} and generates the output current I_{OUT} based on the formula $I_{OUT} = k \cdot I_{REF}$, where k denotes an amplification (or mirroring) coefficient. **Figure 6** depicts the BD configuration of a simple CM. Obviously, more complicated CM structures can be designed using bulk-driven transistors [7].

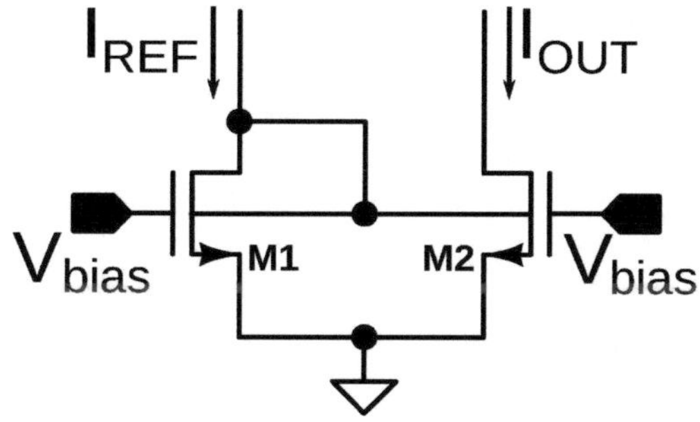

Figure 6.
Simple BD current mirror.

Bulk terminals of both MOS devices M1 and M2 are tied together and connected to the input branch. The gate terminals are biased by static voltage V_{bias}. On the input side, the voltage drop V_{BS} is created by the input reference current flow. This voltage is also applied to the output branch, through the bulk terminal of M2. Hence, the output current is modulated by means of bulk-driving according Eq. (3).

3.2 BD differential amplifier

Another widely and frequently implemented circuit topologies the differential amplifier is depicted in **Figure** 7; however, in the bulk-driven configuration. The devices M1 and M2 have their gate terminals tied to the lowest potential to guarantee the highest possible level of inversion. The traditional topology of the differential amplifier suffers from a limited input common-mode range (V_{CM}) due to necessity of exceeding the input pair threshold voltage and minimal saturation voltage of the biasing high-side transistor M_b. The V_{CM} voltage range can be described by Eq. (13).

$$V_{CM} = V_{DD} - V_{DSsat(Mb)} - V_{TH} \tag{13}$$

The input BD transistors are used to obtain the rail-to-rail input voltage range, which is important for achieving a sufficient voltage swing when low supply voltage value is used, which greatly enhances the input common-mode range (ICMR). Additional benefit of employing the bulk-driven differential amplifier rather than conventional one lies in highly linear voltage-to-current conversion thanks to

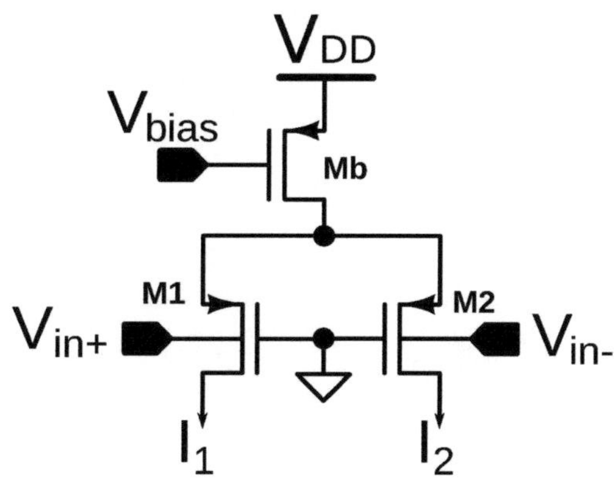

Figure 7.
BD differential pair.

almost perfectly constant transconductance g_{mb}. The disadvantage of presented topology is linked to the grounded gate terminals of the differential transistor pair. This way, the ground noise will be picked up by the transistors, which degrades the power supply rejection ratio (PSRR) [8]. An example of the BD differential pair use was published in [9], where it was used as the input pair of a variable-gain amplifier (VGA).

3.3 Variable-gain amplifier (VGA)

3.3.1 Description of BD VGA

Figure 8 shows the block diagram of a two-stage VGA. The first stage is formed by a variable-gain differential difference amplifier (DDA) designed using BD approach. The second stage has a fixed gain and is created by a BD common-source amplifier (CSA). For stabilization of the operational point of both stages, two BD common-mode feedback (CMFB) circuits have to be employed. To achieve good stability of the CMFB loop as well as the whole two-stage VGA, frequency compensation circuitry has been applied.

The schematic diagram of the low-voltage VGA circuit is depicted in **Figure 9**. The input stage of the proposed topology is formed by DDA with bulk-driven MOS transistors, in order to obtain rail-to-rail input voltage range. The negative aspect of this solution lies in the reduced voltage gain and the gain-bandwidth product (GBW) [8]. Therefore, it is safe to state, that the proposed approach is suitable for low-voltage and low-frequency applications.

In general, the overall VGA voltage gain can be controlled by adjusting either the total conductance or the total output impedance [8]. Thus, transistors M5 and M6 were employed to control the VGA gain. Modification of control voltage V_{CTRL} modulates the current flow through the input devices M1 and M2, which eventually changes their effective transconductance. This will, in effect, vary the voltage amplification of the first stage, which will result in modification of overall gain of the proposed VGA. The total VGA voltage gain is directly proportional to the transconductance of the gain control devices g_{m5} and g_{m6}. The input transistors along with the gain control devices form a cascode configuration, hence their small-signal parameters affect the overall output resistance of the first stage. Detailed small-signal analysis of the complete VGA is conducted in the following subsection. The second stage represents a BD common-source stage with the fixed gain that offers rail-to-rail input voltage range and a wide swing (nearly rail-to-rail) at the VGA output. Two CMFB circuits were used to achieve the stable operational point of both VGA stages.

Figure 9 depicts the implemented frequency compensation circuit. The second amplification stage of the proposed VGA consists of devices M9–M12 and transistors M13–M16 along with capacitors C_c are responsible for the stability of the whole

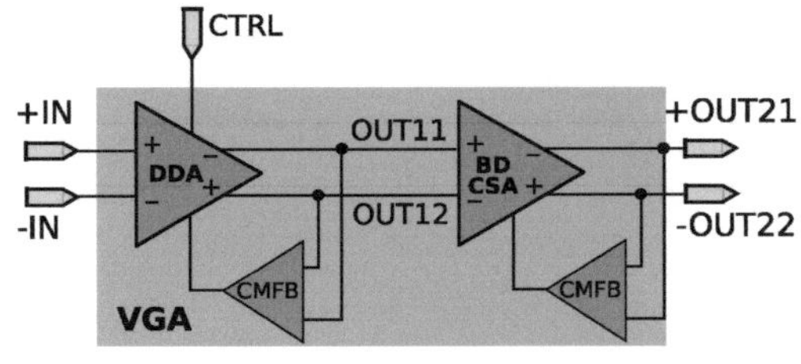

Figure 8.
Block diagram of VGA.

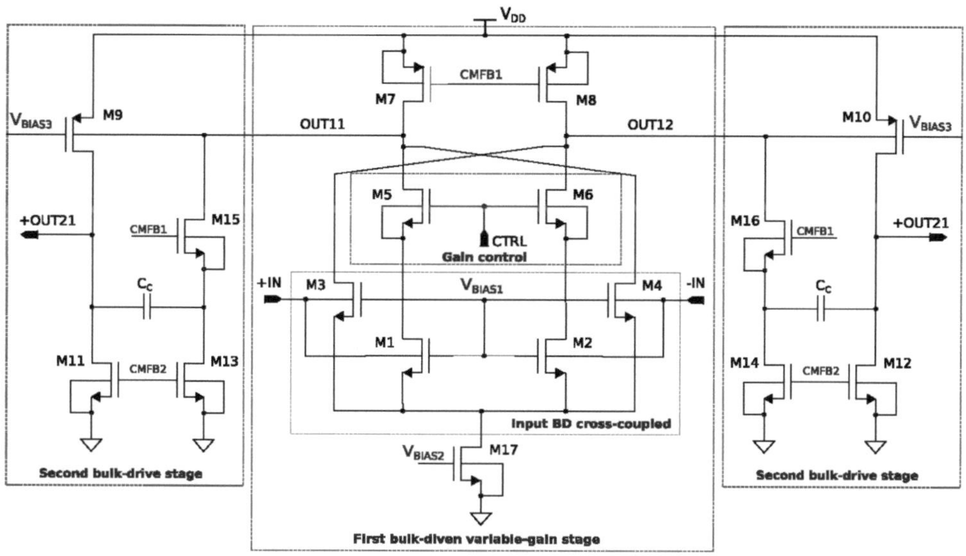

Figure 9.
Schematic diagram of VGA.

circuitry. The proposed compensation approach removes the feed-forward path between the outputs of both amplifying stages by introducing a sufficient voltage gain the feedback path (transistor M15 and M16). The position of the dominant pole is, therefore, maintained, but the position of the output pole gets shifted by the factor of introduced gain. The compensation capacitor C_c is employed between the virtual ground handled by the devices M13 and M14, and the output node of the VGA.

3.3.2 Design principle of VGA based on small-signal analysis

From the design point of view, it is important to ensure good stability and investigate the parameters that influence the gain of the proposed two-stage VGA. For this purpose, the small-signal model of VGA shown in **Figure 10** was used. We have to note that the output capacitance of the first stage was neglected. Thanks to the overall symmetry of the topology, we can perform the small-signal analysis just for one half-circuit and investigate the influence of $+V_{in}$ and $-V_{in}$ separately. The total voltage gain will be, therefore, equal to the sum of the partial contributions of the respective gain stages.

The resulting formula defining the total DC gain of the discussed VGA is follows:

$$A_v = G_{m1}R_{out1} \cdot G_{m2}R_{out2} = A_1 \cdot A_2 \tag{14}$$

where A_1 and A_2 is gain of the first stage and the second stage, respectively. The total transconductance and the total output resistance of the first VGA stage are denoted as G_{m1} and R_{out1}, respectively. The same parameters associated with the second stage are named G_{m2} and R_{out2}. The impact of frequency compensation block has been neglected for additional simplification of the small-signal analysis Eq. (14).

G_{m1} and G_{m2} depicted in **Figure 10** are defined follows:

$$G_{m1} = g_{mb1} \cdot \frac{g_{m5}}{g_{ds1} + g_{m5}} - g_{mb4} = g_{mb1} \cdot C - g_{mb4} \tag{15}$$

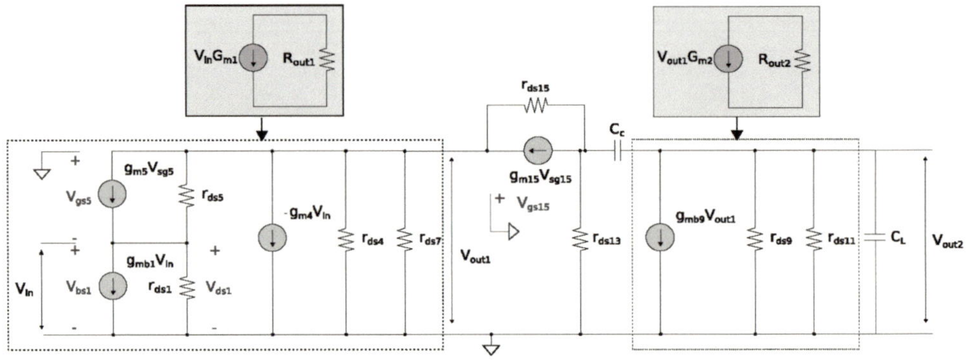

Figure 10.
Small-signal model of the VGA.

$$G_{m2} = g_{mb9} \tag{16}$$

where g_{mb1}, g_{mb4}, and g_{mb9} are bulk transconductance of transistors M1, M4, and M9, g_{m5} is a transconductance of transistor M5, while g_{ds1} and g_{ds5} are output transconductance of transistors M1 and M5, respectively. In Eq. (15), we consider that $g_{m5} > > g_{ds5}$. The total transconductance of the first stage can be varied by the transconductance of transistor M5, while the second stage has a fixed value of transconductance.

$$C = \frac{g_{m5}}{g_{ds1} + g_{m5}} = \frac{1}{1 + \frac{g}{g_{m5}}} \tag{17}$$

One can observe that the total transconductance of VGA (G_{m1}) can be controlled by transconductance g_{m5}, which depends on the control voltage (CTRL). **Figure 11** shows the dependence of the total transconductance G_{m1} on the coefficient C that can vary in the range from 0 to 1. If $g_{m5} < < g_{ds1}$, the coefficient C from Eq. (17) becomes zero and the total transconductance of VGA will be equal to $-g_{mb3}$. In the opposite case (when $g_{m5} > > g_{ds1}$), coefficient C is equal to 1 and G_{m1} will be equal to $g_{mb1} - g_{mb3}$. This means that by increasing the transconductance g_{m5}, the total transconductance G_{m1} is decreased.

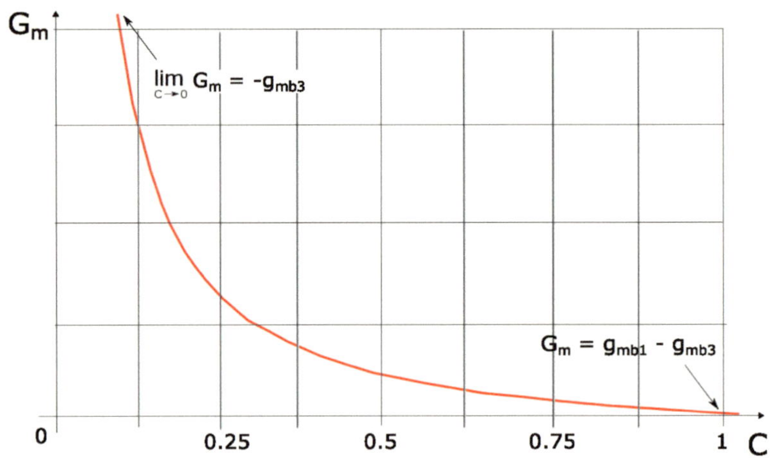

Figure 11.
Transconductance G_{m1} versus the coefficient C.

Besides, the total transconductance G_{m1} also depends on g_{ds1}/g_{m5} ratio. If numerator and denominator of the ratio are divided by I_{ds1}, the following expression can be written:

$$\frac{g_{ds1}}{g_{m5}} = \frac{\frac{g_{ds1}}{I_{ds1}}}{\frac{g_{m5}}{I_{ds1}}} = \frac{V_{CTRL} - V_{TH1}}{2 \cdot (V_{A1} + V_{ds1})}, \tag{18}$$

where V_{CTRL} is the control voltage of VGA, and V_{TH1}, V_{A1}, and V_{ds1} is the threshold voltage, Early voltage and voltage between source and drain of transistor M1, respectively. From Eq. (18), it can be observed that g_{ds1}/g_{m5} ratio, which controls the total transconductance G_m, depends on the control voltage of the proposed VGA. Using the small-signal model depicted in **Figure 10**, the output resistances R_{out1} and R_{out2} of the proposed VGA are expressed as

$$R_{out1} = \underbrace{\left[(1 + g_{m5} r_{ds1}) r_{ds5} + r_{ds1}\right]}_{\text{A}} \| \underbrace{r_{ds4} \| r_{ds7}}_{\text{B}}, \tag{19}$$

$$R_{out2} = r_{ds9} \| r_{ds11}, \tag{20}$$

where r_{ds1}, r_{ds4}, r_{ds5}, r_{ds7}, r_{ds9}, and r_{ds11} are the output resistances of respective devices. The results of the presented small-signal analysis indicate that the first stage gain can be controlled by varying the transconductance of device M5, which is linearly dependent on the VGA control voltage (CTRL). However, the gain of the second stage defined by g_{mb9}, r_{ds9}, and r_{ds11}, is completely independent of CTRL.

The known location of the poles and zeros present in the proposed topology and their analytical definition is crucial for stability evaluation. The small-signal model of the circuit (**Figure 10**) has been used again for the pole-zero analysis. The overall small-signal model for both stages of the discussed VGA can be simplified to a single voltage-controlled current source (VCCS) with their associated output resistances R_{out1} and R_{out2}, respectively (dotted line and gray boxes). The respective small-signal current generation can be expressed $V_{in}G_{m1}$ and $V_{out1}G_{m2}$. Moreover, adopting the simplified model, the low-frequency gain can be re-written in more detail as follows:

$$A_v = \frac{A_1 A_2 \cdot \left[A_{M15} r_{ds13} + r_{ds13} + r_{ds15}\right]}{A_{M15} r_{ds13} + R_{out1} + r_{ds13} + r_{ds15}} \approx A_1 A_2, \tag{21}$$

where A_{M15} is a small-signal gain of transistor M15 while r_{ds13} and r_{ds15} are the output resistances of transistors M13 and M15, respectively. If we consider that $R_{out1} < < A_{M15} r_{ds13} + r_{ds13} + r_{ds15}$, the low-frequency gain of the designed amplifier can be approximately expressed as $A_1 A_2$. Using this approximation, the error between simulated and calculated low-frequency gain will be about 3%.

The first pole defines the amplifier bandwidth (BW), and it is approximately given by Eq. (22).

$$p_1 \approx \frac{-1}{(C_L + C_c) R_{out2} + \frac{A_{M15} A_2 R_{out1} r_{ds13} C_c}{A_{M15} r_{ds13} + r_{ds15}}} \tag{22}$$

It was considered in the approximation that R_{out1} and r_{ds13} are much lower than $A_{M15} r_{ds13} + r_{ds15}$. Since, the VGA has the second stage with a fixed gain, by observation of Eq. (22), it can be concluded that BW might be influenced by resistance R_{out1}. Therefore, the design of the discussed VGA can be conducted in two different ways.

The resistance R_{out1} is dependent on coefficients A and B (Eq. (19)), thus the bandwidth of the VGA can be either influenced by the controlling voltage or not. The resulting R_{out1} is defined by parallel combination of both coefficients, therefore their values determine the frequency properties of R_{out1}. When the portion of A dominates (A < < B), the control voltage will adjust G_{m1} and R_{out1}, which eventually allows for greater range of gain control. On the other hand, the bandwidth will be also influenced. The other option is when R_{out1} is determined by B (A > > B). In this case, the gain can be controlled in narrower range, but the bandwidth will be independent of CTRL. Based on the presented design considerations, we have chosen to employ the second option, in order to obtain constant bandwidth across the control voltage range.

To achieve a stable operation of the proposed VGA topology, the so-called *pole splitting* compensation technique has been implemented. The second pole p_2 needs to be positioned at much higher frequency then the GBW. The analytical definition of p_2 to GBW ratio is expressed as follows:

$$\frac{p_2}{GBW} = \frac{1/p_1^2 \cdot [A_{M15} r_{ds13} + r_{ds15} + R_{out1}]}{A_1 A_2 R_{out2} r_{ds13} r_{ds15} C_L C_c \cdot [r_{ds15} + R_{out1}]} \tag{23}$$

It is obvious, that for satisfactory amount of phase margin (PM), the numerator of Eq. (23) has to be greater than the denominator. It is widely accepted that PM = 60° is sufficient for maintaining a stable operation of amplifier circuits, which will be met when the ratio $p_2/GBW \approx 2$. An important remark regarding PM is that it is negatively affected by gain A_1, hence very high value of A_1 will result in lowered PM. On the other hand, if the VGA is stable even with high gain level A_1, it will maintain stability for low A_1, too [8].

3.4 Low-voltage BD comparator

In this section, we would like to describe the proposed topology of a voltage comparator operating with the power supply voltage (V_{DD}) of 0.4 V. Due to very limited voltage span, the comparator has to be able to process the input signals in rail-to-rail range. Another important feature regarding voltage comparators is the hysteresis. The proposed comparator topology contains four levels of hysteresis programmable by input digital code. Additional but very welcome feature especially in low-power circuits is enable function. Its whole purpose is to inhibit the circuit function and lower the internal current consumption into the leakage range. One can easily observe the main advantages of the discussed comparator topology. Unlike the traditional comparator circuit with three stacked MOS transistors, the proposed one, employing just two stacked devices, can obviously function with lower supply voltage levels. The second benefit lies in the rail-to-rail input voltage range achieved by only one differential structure, rather than parallel combination of two separate PMOS and NMOS amplifiers. Another important contribution, if compared to the traditional comparator, is based on the lack of the internal bias reference block. This feature simplifies and speeds up the design procedure, lowers the overall quiescent current consumption and enables setting the current flow through the circuit branches just by respective transistor sizing.

3.4.1 Description of low-voltage comparator

The analog core of the proposed circuit is depicted in **Figure 12** along with the devices' dimensions. Let us discuss the topology of the ultra-low-voltage rail-to-rail

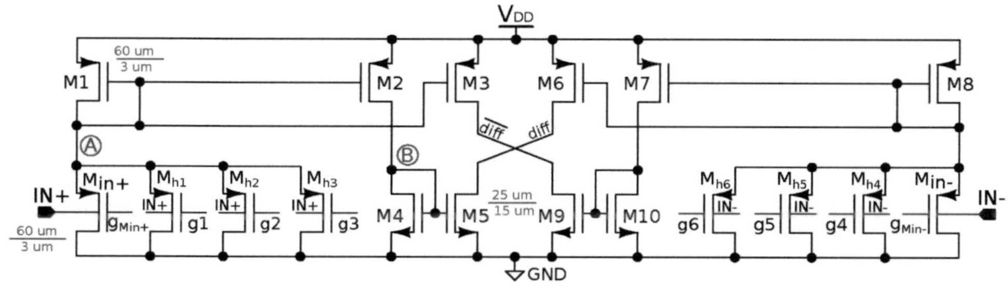

Figure 12.
Analog core of ultra-low-voltage comparator.

voltage comparator. As one can observe, the input signal is processed by bulk-driven MOS transistors. This is an elegant way to solve the issues with the threshold voltage of respective devices. The most dreaded problem associated with rather exotic bulk-driven circuits is so-called *latch-up effect*, which causes catastrophic short between the power supply rails. However, with power supply voltage set to V_{DD} = 0.4 V, such an event cannot occur, because the threshold voltage of PN junction even at T = 150°C does not decrease below the specified value.

The input PMOS devices M_{in+} and M_{in-} act as a current sources. The static current flowing through the input branches is being modulated by the input voltage according to Eq. (3). This voltage-to-current conversion occurs in highly linear manner thanks to almost ideally constant transconductance $g_{mb} = \frac{\partial I_D}{\partial V_{BS}}$ across whole input voltage range. The current from the input branches is afterward copied by means of unity current mirror system comprised by devices M1–M10. The differential voltage is created in nodes called *diff* and $\overline{diff}$ by current-to-voltage conversion. The differential signal is further processed by digital block, which is depicted in **Figure 13**.

The gate terminals of low-side devices employed in analog core are controlled by the digital part. The input devices can be cut-off when their gate terminals are pulled up—digital block issues logic one, when the circuit function is being inhibited. Otherwise, the gates are pulled down by logic zero, which sets their operation in to active, saturation, and region. The schematic diagram omits four

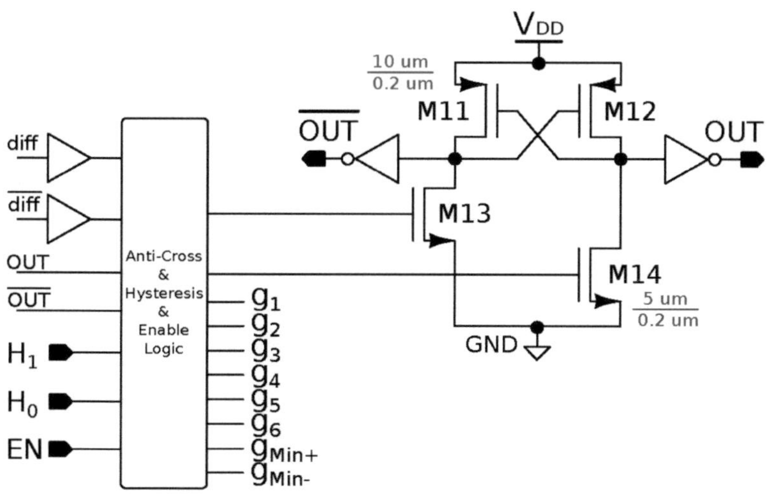

Figure 13.
Digital control and output block of the ultra-low-voltage comparator.

pull-up and pull-down devices, which are responsible for ensuring a definite and known potential in the analog core nodes, when the circuit function is being disabled.

Transistors M_{h1}–M_{h6} are used to introduce hysteresis into the transfer characteristics without any external component required. The principle of operation is rather simple, yet very effective. In our case, we created four different levels of symmetrical hysteresis. The amount of hysteresis is programmable by 2-bit input code processed by the controlling digital block. Based on this information, the gate terminals of respective devices are pulled up or down and creates hysteresis of ± 0, 10, 20, and 50 mV. Therefore, the static current flow in the input branches is disbalanced, which in effect, shifts the trip point of the comparator. The beauty of this method is in minimal area and current consumption overhead, since the amount of the hysteresis is determined by transistor size. Another important feature is the possibility of creating customized shape (e.g., non-symmetrical) and level of hysteresis.

The controlling block and output latch are depicted in **Figure 13**. The controlling block processes the input signals of enable and hysteresis functions and the feedback signal from the output latch and the differential voltage from the analog core. The combinatorial logic issues logic states for the gate terminals of transistors in the analog core accordingly. It is also responsible for protecting the output latch from simultaneous switching based on the feedback information. Such an event has been observed with highly rippled power supply voltage. The output signal from the latch is afterwards "buffered" by digital inverter(s) sized accordingly to the capacitance load and speed/slew-rate requirements.

3.4.2 Design considerations and analysis

Let us discuss the design considerations of the proposed comparator topology. Since the power supply voltage has already been set to V_{DD} = 0.4 V due to bulk-driven input stage, we need to determine the operational temperature, or in other words the temperature corners. The usual choice is set within the industrial standard −20°C and 85°C, which was also our case. Another important aspect is a selection of CMOS fabrication process node. Current standard (as of 2020) for analog/mixed-signal designs presets the choice into planar deep sub-micron technologies (e.g., 500, 350, 130, or 90 nm) thanks to their fair trade-off between the cost and the general properties of the process and available electronic components. Analog/mixed-signal designs in finer technology nodes usually face complications requiring advanced solutions and workarounds, which outweighs the advantages linked to smaller silicon circuit area or higher operational speed.

The final transistor sizes can be seen in **Figures 12** and **13**. From the designer's point of view, the analog part has to be symmetrical, which decreases the overall effort required during the design period. One can observe, rather large MOS transistor dimensions. This choice has been done, in order to mitigate the process fluctuations associated with nanoscale CMOS processes. Furthermore, it also increases the level of inversion within the respective devices since higher the inversion coefficient, the lower current mismatch and process sensitivity.

The small-signal model of the proposed topology is shown in **Figure 14**. Thanks to the overall symmetry of the analog core, we will greatly benefit from analyzing only one half of the circuit. We have chosen the left-hand side, for our analysis. Furthermore, the devices from both topology sides are mutually interchangeable for the sake of small-signal analysis, if needed. The nodes marked "A" and "B" in the transistor-level schematic (**Figure 12**) are also depicted in the small-signal model. Another analysis simplification is, that the overall voltage amplification can be

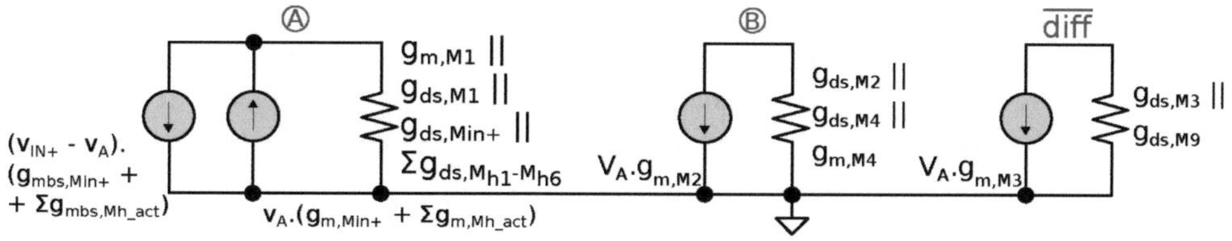

Figure 14.
Small-signal model of the proposed ultra-low-voltage comparator.

expressed as a product of partial voltage amplifications (or as a sum of partial voltage gains) of respective stages. For even further simplification, we have also omitted the enable pull-up and pull-down devices, which do not affect the overall accuracy that much and can be therefore neglected.

The first stage is comprised of devices M_{in+}, M_{h1--6} and diode-connected MOS transistor M1. The input node is the actual comparator input terminal. The output node, in this case, is the node named "A". As one can observe, the bulk-driven M_{in+} inherently contains a negative voltage feedback, which unfortunately decreases the voltage amplification below unity. This is depicted as two current sources working against each other in the small-signal model. The precise analytical expression of the first stage voltage amplification is defined by Eq. (24). It is apparent, that the overall amplification is dependent on the activated hysteresis devices.

$$A_{V_A} = \frac{g_{mbs_{Min+}} + \sum g_{mbs_{Mh\_act}}}{g_{ds_{Min+}} + g_{m_{Min+}} + g_{mbs_{Min+}} + g_{ds_{M1}} + g_{m_{M1}} + \sum g_{ds_{Mh1-h6}} + \sum g_{m_{Mh\_act}}} \tag{24}$$

The second amplification stage is consist of devices M2 and M4. The input port is the "A" node and the output port would be the node named "B." The devices form a classic configuration of PMOS common-source amplifier with diode-connected transistor acting as a load. The small-signal model yields an Eq. (25), which defines the voltage amplification of the second stage.

$$A_{V_B} = \frac{g_{m_{M2}}}{g_{ds_{M2}} + g_{ds_{M4}} + g_{m_{M4}}} \tag{25}$$

The third and final stage of the model is comprised of devices M3 and M9. However, due to already explained reasons, we can swap the MOS transistor M9 with device M5, if desired. The topology of this stage can be described as a common-source amplifier with MOS transistor acting as load with fixed-bias. The small-signal model also reveals the fact, that the final stage is driven by the first stage only. Hence, the second stage does not contribute to the overall amplification, at all. The analytical voltage amplification provided by the final stage is defined by Eq. (26).

$$A_{V_{\overline{diff}}} = \frac{g_{m_{M3}}}{g_{ds_{M3}} + g_{ds_{M9}}} \tag{26}$$

As we discussed earlier, the final voltage amplification is a product of partial amplification contributions of the respective stages. However, the final analytical formula is unnecessarily complex and can be significantly simplified, if the circuit designer follows some of the common sense rules. The most basic one is associated with the "reasonable" channel length and/or minimized channel length modulation along with saturation operation mode of the transistors employed in the analog core. If this requirement is met, the sum of output conductance g_{ds} of respective devices in the denominator of Eq. (24) can be neglected, since it is several orders of magnitude lower than the sum of their transconductances g_m. Keeping in mind that the second stage does not contribute to the overall gain, we can re-write the resulting formula as follows:

$$A_V \approx \left(\frac{g_{mbs_{Min+}} + \sum g_{mbs_{Mh\_act}}}{g_{m_{Min+}} + g_{mbs_{Min+}} + g_{m_{M1}} + \sum g_{m_{Mh\_act}}} \right) \cdot \left(\frac{g_{m_{M3}}}{g_{ds_{M3}} + g_{ds_{M9}}} \right) \tag{27}$$

It is apparent that for the maximized voltage gain, one should increase the transconductance of M3 and minimize the output conductance of devices M3 and M9. Since the topology is symmetrical, the same applies for the counterparts of discussed transistors.

Another important design consideration is the minimum power supply voltage the circuit can reliably operate with. The presented topology contains only two stacked transistors that need to be saturated for correct functioning. Therefore, we can express the *theoretical* minimum power supply voltage at room temperature in as follows:

$$V_{DD_{min}} = 2 \cdot V_{DS_{sat}} \approx 2 \cdot \left(4 \cdot \frac{k.T}{q} \right)_{T=300K} \approx 206.88\, mV \tag{28}$$

The expression is taken from the EKV MOS transistor model theory for deep weak inversion [4]. In real scenario, the saturation voltage dependence on the level of inversion is of square root trend. Furthermore, considering a non-ideal conditions, the more realistic expectation is about $V_{DD_{min}}$ = 230 mV at room temperature.

3.5 Ultra-low-voltage BD charge pump

3.5.1 Description of the BD charge pump

CMOS realization of the conventional charge pump based on the cross-coupled voltage doubler is shown in **Figure 15**. This charge pump represents one of the highly efficient topologies suitable for on-chip implementation. Both types of MOS devices are employed in the cross-coupled charge pump. The proposed control approach mitigates the excessive voltage stress, which the thin gate oxide of the MOS transistor is exposed to. The voltage drop is comprised of the equivalent resistance of the switched capacitors and the voltage drop across the turned-on transistor [10].

The cross-coupled charge pump (**Figure 15**) is based on two inverters connected in cross-couple fashion. The discussed topology, exhibits the same level of minimal input switching voltage $V_{\phi min}$ as the traditional CMOS inverter the sum of threshold voltages of the respective MOS transistors employed. Hence, a specific modifications to the circuit topology are required, in order to enable the implementation in ultra-low-voltage applications.

One of the possible alternatives is to employ an inverter with resistor acting as a load, which would create the so-called pseudo-inverter. However, the solution is unacceptable in low-power systems, since it notably increases the internal current consumption of the charge pump. Another possibility is to tie the bulk terminal of the MOS transistors to a fixed potential and lower their threshold voltage (**Figure 16(b)**). Unfortunately, this approach substantially increases the subthreshold leakage of the MOS transistors and their bulk current, as well. Hence, it

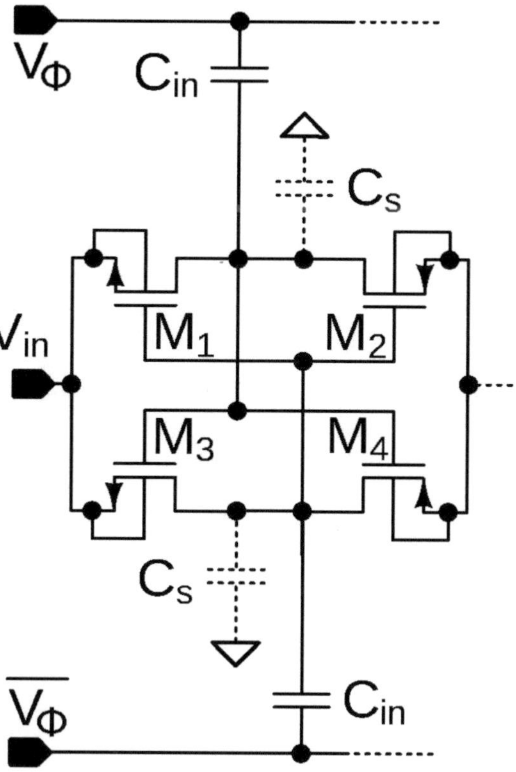

Figure 15.
Conventional cross-coupled CP (one stage).

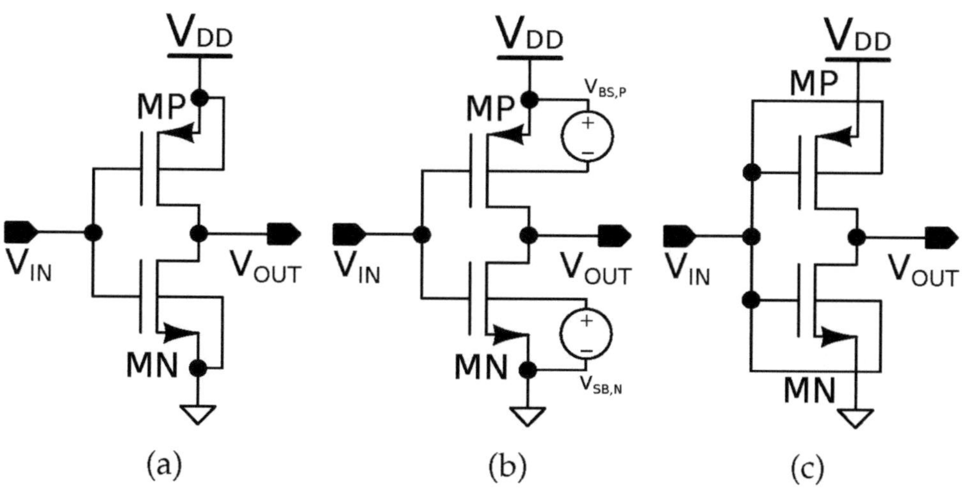

Figure 16.
Different topology inverters in CMOS technology.(a) CMOS inverter, (b) CMOS inverter: reduced VTH (c) CMOS inverter: dynamic VTH.

suffers from the same drawback as the solution mentioned previously. The most effective approach, as it seems, is the so-called dynamic threshold voltage control. The basic idea is depicted in **Figure 16(c)**. As one can observe, the bulk terminals are connected to the input signal. This way, the cut-off transistor exhibits nominal threshold voltage and the turned-on transistor exhibits lowered threshold voltage

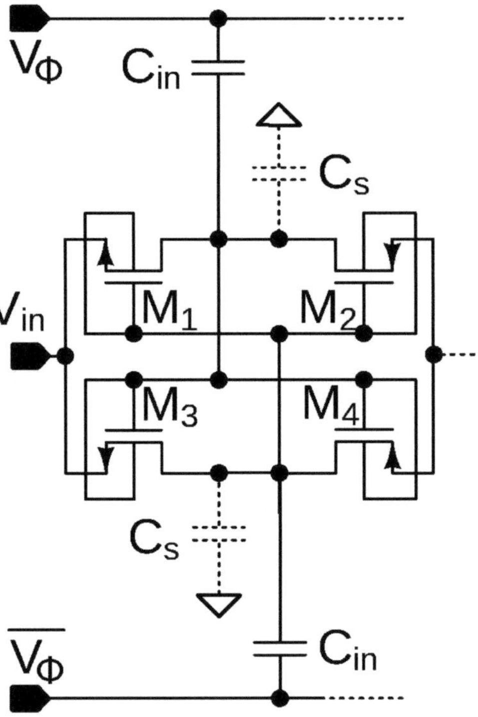

Figure 17.
Bulk-driven cross-coupled CP (one stage).

level, which improves its drain current and switching speed. The trade-off of this method is increased leakage current of one transistor only.

Dynamic threshold inverter topology (**Figure 16(c)**) can be considered as an appropriate building cell for ultra-low-power and low-voltage charge pump designs. A very important design consideration for such a system is inevitability of twin-well CMOS fabrication process, since both PMOS and NMOS devices have to be isolated from the common substrate by their own wells. Another consideration is restricted power supply voltage. Its level must not exceed 0.6 V (at room temperature). Otherwise, parasitic NPN and PNP bipolar transistors will turn on, which with the high probability may cause triggering of latch-up effect. These limitations have to be taken into account, if designing the charge pump with dynamic V_{TH} MOS inverters.

The BD charge pump based on the cross-coupled topology is shown in **Figure 17**. The charge pump employing dynamic threshold CMOS inverter is able to process very low levels of input voltages. This also represents the main advantage of the proposed topology [10]. However, the drawback lies in the limited voltage range the charge pump can work with, because of the increased risk of latch-up triggering. Hence, the main application area of such charge pump topology is constrained to low-voltage and low-power systems.

3.5.2 Design procedure of BD charge pump

Theoretical value of the output voltage can be expressed as follows:

$$V_{out} = V_{in} + N\left(V_{\phi}.\frac{C_{in}}{C_{tot}} - I_L\left(R_{on} + \frac{R_{eq}}{2}\right)\right), \quad (29)$$

where N is a number of stages used, V_{in} is the input voltage, V_{out} is the output voltage, V_{ϕ} is the switching voltage, C_{in} is the switching capacitance, C_{tot} is sum of switching capacitance and parasitic capacitance $C_{tot} = C_{in} + C_s$, I_L is sum of output and leakage currents $I_L = I_{out} + I_{leakage}$, R_{on} is switch-on resistance of NMOS and PMOS transistor, and R_{eq} is equivalent resistance of switching capacitor.

Due to limited voltage headroom, the sub-threshold operation is usually chosen for low-voltage/low-power charge pump designs. Although, when the input voltage is lesser than threshold voltage of MOS transistors, the overall efficiency will be degraded, since the switching properties of the devices are drastically reduced. The dependence of MOS transistor turn-on resistance R_{ON} on the overdrive voltage (V_{EFF} or $V_{GS} - V_{TH}$) is defined in Eq. (30).

$$R_{ON} \sim \frac{1}{e^{V_{GS}-V_{TH}}} \tag{30}$$

It can be observed that effective voltage V_{EFF} in relation to R_{ON} acts as argument of exponential function located in denominator, which implicates the substantial sensitivity to change of this quantity. In order to overcome this issue, many techniques such as dynamic control, feed-forward biasing of the bulk electrode, multi-stage topology, implementation of charge transfer switches (CTS) parallel with pass gate switches (PGS) or backward control of the previous stage by body electrode voltage can be included in the low-voltage CPs design [11–14].

3.6 Evaluation of the proposed LV circuits

In order to fully verify the design concepts discussed in this chapter, the proposed circuit topologies have been designed in general purpose twin-well 130 nm CMOS technology. Both types of fabricated long-channel devices exhibit the standard threshold voltage around V_{TH} = 260 mV at room temperature. **Figure 18** depicts the micrograph of the fabricated prototype chip along with the physical layout design. For better clarity, we omitted some of the top metal layers. The blue rectangles pin-point the location of the proposed circuits on chip. The chip also contains a stand-alone test transistor in order to verify the devices' compact simulation model accuracy and fabrication process fluctuations as well.

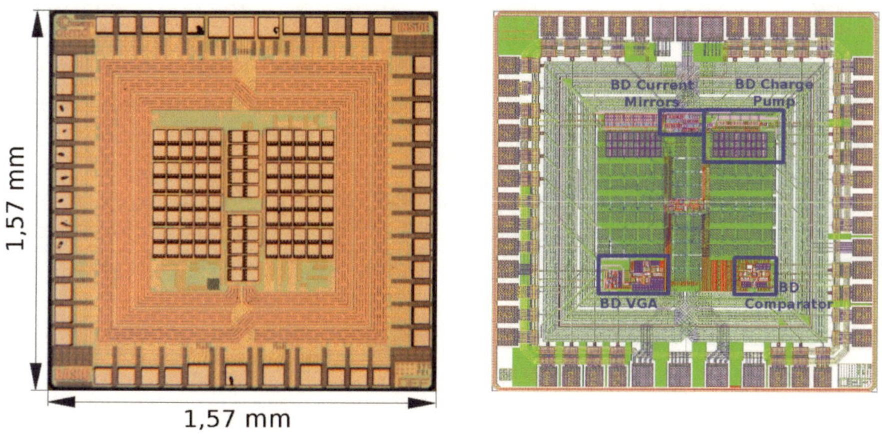

Figure 18.
Micrograph and physical design of the prototype chip.

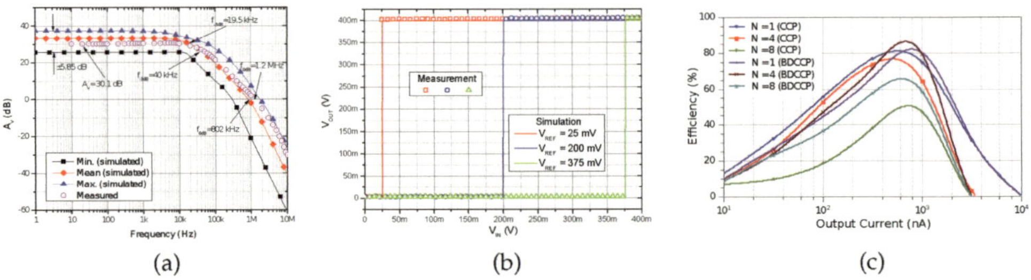

Figure 19.
Measured and simulated parameters of the proposed circuit topologies on the prototype chip. (a) VGA frequency response, (b) Comparator transfer characteristics, (c) Charge-Pump efficiency.

The selected parameters of the proposed VGA, voltage comparator, and charge pump are depicted in **Figure 19**. The graphs contain the measured and simulated results for direct comparison and evaluation.

The comparison of simulated and measured data of frequency response of the proposed VGA is depicted in **Figure 19(a)**. We used the Monte-Carlo analysis results to obtain the borders of the expected gain range. As one can observe, the measured frequency response remains between the borders specified by the simulation and deviates only slightly from the Monte-Carlo mean curve. The measurement and simulation conditions were identical, the load capacitance of 10 pF, the control voltage V_{CTRL} = 0.1 V, ambient room temperature, and power supply voltage of V_{DD} = 0.6 V were used. The obtained measured results at these operating conditions were the following. The voltage gain of a_V = 30 dB, gain-bandwidth of GBW = 1.2 MHz and bandwidth off_{-3dB} = 40 kHz were observed. The worst-case discrepancy of the measured low-frequency gain compared to the mean Monte-Carlo curve is approximately ±5.85 dB.

The transfer characteristics of the proposed rail-to-rail comparator are depicted in **Figure 19(b)**. The simulated results correlate very well with measured curves for all input voltage conditions. In our experiments, the comparator exhibited correct function for all four levels of hysteresis and the rail-to-rail operation has been confirmed by setting the reference voltage only 3 mV from the power supply range. This test can be considered quite strict, since it would also reveal issues with the input offset voltage. Monte-Carlo simulations performed on 3000 samples in corner and ambient temperatures resulted in the mean value of input offset V_{offset} = 592 uV with standard deviation of σ = 1.91 mV [15]. Furthermore, the power consumption at V_{DD} = 0.4 V has been measured in the upper half of nW range including the ESD structures leakage, PCB, and oscilloscope probe parasitics. However, we have also observed correct operation with V_{DD} = 0.25 V, which indeed astonishing result.

Figure 19(c) shows dependence of the efficiency on the output current. This is the example where the comparison of different charge pumps based on GD and BD cross-coupled inverter were compared. The best efficiency for the given parameters can be observed for the output current of $1\mu A$, where the BD cross-coupled charge pump achieves the efficiency of 80% in four-stage architecture (N = 4) in given CMOS technology.

4. Conclusions

Considering the current onset of ultra-low-voltage and ultra-low-power operation requirements for today's CMOS analog/mixed-signal ICs and their fabrication, a promising alternative to standard design approach and circuit topologies are discussed in this chapter. We presented low-voltage and low-power design techniques, which are focused on driving the MOS transistor through its bulk terminal, as well as setting the operating point of employed MOS transistors within the sub-threshold (weak inversion) or moderate inversion region. The combination of described design techniques provides the significant power consumption minimization (nW feasible), while maintaining acceptable circuit performance and parameters. The motivation for research and development in given scientific field is enormous and new published results are expected to grow in upcoming years.

The proposed circuit topologies of basic analog IC building blocks have been designed and fabricated in 130 nm twin-well general purpose CMOS technology with industrial operating temperature range taken into account. The experimental measurements performed on the prototype chip samples confirm a successful implementation and correct circuit operation with ultra-low-power supply voltage. Hence, we can state that the feasibility of presented IC design approach has been successfully demonstrated and the circuit topologies have been silicon-proven, which opens the door for even deeper gradual investigation and understanding of given scientific topic with promising future impact on the IC industry as well.

Author details

Daniel Arbet†, Lukas Nagy† and Viera Stopjakova*†
Faculty of Electrical Engineering and Information Technology, Slovak University of Technology, Bratislava, Slovakia

*Address all correspondence to: viera.stopjakova@stuba.sk

† These authors contributed equally.

References

[1] Stopjakova V, Kovac M, Arbet D, Nagy L. Towards energy-autonomous integrated systems through ultra-low voltage analog IC design. In: 26th International Conference "Mixed Design of Integrated Circuits and Systems"; 2019. pp. 38-45

[2] Rakus M, Stopjakova V, Arbet D. Design techniques for low-voltage analog integrated circuits. Journal of Electrical Engineering. 2017;**68**(4):245-255

[3] Tajalli A, Leblebici Y. Extreme Low-Power Mixed Signal IC Design: Subthreshold Source-Coupled Circuits. New York: Springer-Verlag; 2010

[4] Binkley D. Tradeoffs and Optimization in Analog CMOS Design. UK: Wiley; 2007

[5] Enz C, Krummenacher F, Vittoz E. An Analytical MOS Transistor Model Valid in all Regions of Operation and Dedicated to Low-Voltage and Low-Current Applications. Switzerland: Springer; 1995

[6] Silveira F, Flandre D, Jespers PGA. A gm/ID based methodology for the design of CMOS analog circuits and its application to the synthesis of a silicon-on-insulator micropower OTA. IEEE Journal of Solid-State Circuits. 1996;**31**: 1314-1319

[7] Rakus M, Stopjakova V, Arbet D. Comparison of gate-driven and bulk-driven current mirror topologies. In: 2016 IEEE 19th International Symposium on Design and Diagnostics of Electronic Circuits Systems (DDECS); 2016. pp. 1-4

[8] Arbet D, Kovac M, Nagy L, Stopjakova V, Sovcik M. Two-stage bulk-driven variable gain amplifier for low-voltage applications. In: 2018 IEEE 21st International Symposium on Design and Diagnostics of Electronic Circuits Systems (DDECS); 2018. pp. 45-50

[9] Sedra AS, Smith KC. Microelectronic Circuits. 6th ed. UK: Oxford University Press; 2011

[10] Nagy G, Arbet D, Stopjakova V, Kovac M. Novel CMOS bulk-driven charge-pump for ultra low input voltage. Radioengineering. 2016;**25**: 321-331

[11] Intaschi L, Bruschi P, Iannaccone G, Dalena F. A 220-mV input, 8.6 step-up voltage conversion ratio, 10.45-uW output power, fully integrated switched-capacitor converter for energy harvesting. In: 2017 IEEE Custom Integrated Circuits Conference (CICC); 2017. pp. 1-4

[12] Ozaki T, Hirose T, Nagai T, Tsubaki K, Kuroki N, Numa M. A 0.21-V minimum input, 73.6% maximum efficiency, fully integrated voltage boost converter with MPPT for low-voltage energy harvesters. In: ESSCIRC 2014—40th European Solid State Circuits Conference (ESSCIRC); 2014. pp. 255-258

[13] Shih Y, Otis BP. An Inductorless DC–DC converter for energy harvesting with a 1.2 uW bandgap-referenced output controller. IEEE Transactions on Circuits and Systems II: Express Briefs. 2011;**58**: 832-836

[14] Peng H, Tang N, Yang Y, Heo DH. CMOS Startup charge pump with body Bias and backward control for energy harvesting step-up converters. IEEE Transactions on Circuits and Systems I: Regular Papers. 2014;**61**:1618-1628

[15] Nagy L, Arbet D, Kovac M, Potocny M, Stopjakova V. Design and performance analysis of ultra-low voltage rail-to-rail comparator in 130 nm CMOS technology. In: 2018 IEEE 21st International Symposium on Design and Diagnostics of Electronic Circuits Systems (DDECS); 2018. pp. 51-54

Section 3

Tunnel Field Effect Transistors

Chapter 3

Tunnel Field Effect Transistors Based on Two-Dimensional Material Van-der-Waals Heterostructures

Jiang Cao

Abstract

The successful isolation of graphene in 2004 has attracted great interest to search for potential applications of this unique material and other newborn members of the two-dimensional (2-D) family in electronics, optoelectronics, spintronics and other fields. Compared to graphene, the 2-D transition metal dichalcogenides (TMDs) have the advantage of being semiconductors, which would allow their use for logic devices. In the past decade, significant developments have been made in this area, where opportunities and challenges co-exist. Stacking different 2-D materials significantly increases the already considerable design space, especially when a type-II band alignment is obtained. This chapter will describe the recent progresses in the tunnel field-effect transistors based on 2-D TMD van-der-Waals heterostructure, which is one of the promising candidates for increasingly important low-power mobile computation applications. Due to their small size, such devices are intrinsically dominated by quantum effects. This requires the adoption of a fairly general theory of transport, such as the nonequilibrium Green's functions (NEGF) formalism, which is a method having been more-and-more used for the simulation of electron transport in nanostructures in recent years.

Keywords: two-dimensional material, van-der-Waals heterostructure, tunnel field-effect transistor, steep-slope switch, subthreshold swing

1. Introduction

The invention of the transistor in 1948 is arguably the major technological break-through of the twentieth century. The transistors are the building blocks of today's microprocessors and computers that are everywhere around us. Nowadays, billions of transistors are integrated into a microchip of only a square centimeter. Since the Nobel prize attributed to Shockley, Brattain and Bardeen in the 1956, and the invention of integrated circuits in the same decade, considerable efforts have been put to keep miniaturizing the metal oxide semiconductor field effect transistors (MOSFETs). A conventional MOSFET structure with descriptions of its working principle are shown in **Figure 1**. From one technology node to the next, MOSFETs are conceived to be smaller (following Moore's law), faster and less power consuming. Thirty years of aggressive scaling have pushed the device dimensions close to the atomic range. The downscaling of MOSFETs has slowed down since the 65 nm node was reached. Issues related to the nanoscale dimensions of the devices started arising.

When the channel length is decreased below 1 μm, additional problems appear and are commonly called short-channel effects (SCEs). The SCEs for the MOSFETs are important when the channel length becomes comparable to the width of the depletion region. When the gate length is scaled down, the gate starts to lose the electrostatic control over the channel. On the other hand the source-drain bias (V_{DS}) gains a larger influence on the barrier. Such an effect is named drain-induced barrier lowering (DIBL). This loss of electrostatic integrity leads to a continue increase of the current and decrease of the off-state potential.

Moreover, the electron mobility is reduced due to collisions with the semiconductor/oxide interface. This surface scattering effect is enhanced by the increase of electric field in the confined regions, which pushes the electrons toward the surface of the device. The reduction of electron mobility is also caused by the necessity of using high doping levels in such scaled MOSFET. Finally, the average velocity of carriers is no longer linearly depend on the electric field in such small devices, which is called the velocity saturation.

The drawbacks of traditional bulk planar transistors have promoted the search for new architectures alternative to MOSFETs. The International Technology Roadmap for Semiconductors (ITRS) [1], which evaluates the technology requirements for the next-generation semiconductor device technology, predicts that additional new materials and transistor geometries will be needed to successfully address the formidable challenges of transistor scaling in the next 15 years. In **Table 1**, some main figures of merit extracted from the ITRS for the short- (2018) and long-term (2026) technologies, both for high-performance and low-power applications, are shown. Since the late 90s, it has been suggested to replace single-gate transistors by multi-gate structures in order to enhance the electrostatic control of the gate. Intel has already switched to the TriGate FET, also known as the FinFET, technology since the 22 nm node. Silicon-on-insulator has also been widely used to improve the performances of transistors, especially decreasing leakage currents [2].

To meet the requirements set by the ITRS for future nodes, scaling down the gate length is critical. The two-dimensional materials (2DMs) provide the ability to control the channel thickness at the atomic level, which will result in improved gate control over the channel and in reduced SCEs. In the next Section, I will discuss the properties of 2DMs and the 2DM van-der-Waals heterostructures (vdWHs) and their numerous possible applications in the electronic devices. Before that, let us

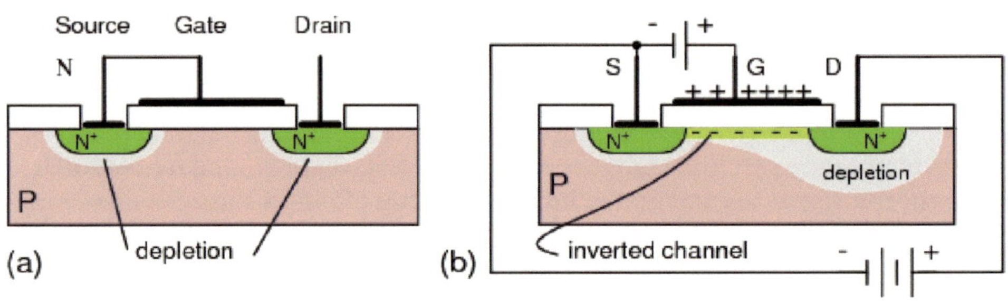

Figure 1.
Schematic cross-section of a N-channel MOSFET: (a) o V gate bias, (b) positive gate bias that charges the gate. The P-type substrate below the gate takes on a negative charge. An inversion region with an excess of electrons forms below the gate oxide. This region connects the source and drain N-type regions, forming a continuous N-region from source to drain.

	HP2018	LP2018	HP2026	LP2026
Channel length (nm)	10.2	11.7	4.9	5.6
V_{DD} (V)	0.78	0.78	0.66	0.66
I_{OFF} (nA/μm)	100	0.01	100	0.04
I_{ON} (μA/μm)	1610	556	1030	337
τ (ps)	0.488	1.564	0.432	1.514

Table 1.
Figures of merit extracted from ITRS [1] for the short- (2018) and long-term (2026) technologies, both for high-performance and low-power applications. V_{DD} is the supply voltage; I_{OFF} and I_{ON} are the drain currents per unit width in the off- and on-state; τ is the intrinsic delay time.

first look at the fundamental power consumption issues in the MOSFET and the FinFET technologies, then one of the promising technologies for solving this problem, namely the tunnel field effect transistors (TFETs).

2. Power consumption issues

Power consumption is a fundamental problem for nanoelectronic circuits. To give some examples, all the smartphones need to be recharged everyday; the data centers in the US used 91 billion kilowatt-hours of electricity in 2013. The power consumption in logic devices closely depends on the supply voltage (V_{DD}) through the following relation.

$$P = \underbrace{\alpha f_c C_L V_{DD}^2}_{\text{operating}} + \underbrace{I_{OFF} V_{DD}}_{\text{stand-by}}, \tag{1}$$

where α is called the activity factor, f_c denotes the clock frequency, C_L is the load capacitance (mostly gate and wire capacitance, but also drain and some source capacitances), and I_{OFF} is the off-state current. In the formula above, we can identify an operating and a stand-by power that both depend on V_{DD}. Lowering V_{DD} is thus necessary to decrease the consumption. However, a strong V_{DD} reduction significantly affects the performances of MOSFETs, as illustrated in **Figure 2(a)**. Indeed, the problem resides in the speed at which the transistor passes from the off- to the on-states as a function of the gate voltage. In the subthreshold regime of MOSFETs, the thermionic effect entails that at least 60 mV are necessary to increase the current by one order of magnitude at room temperature. In other words, the subthreshold swing (SS), i.e. the inverse of the derivative of the subthreshold slope

$$\text{SS} = \left(\frac{\partial \log_{10} I_{DS}}{\partial V_G}\right)^{-1}, \tag{2}$$

has a minimum value of

$$\text{SS}_{\min} = \frac{k_B T}{e} \ln(10) = 60 \text{ mV/dec}, \tag{3}$$

where k_B is the Boltzmann constant, T is the temperature taken at 300 K and e is the absolute value of electron charge. If we keep the same on-current I_{ON} for the transistor while reducing V_{DD}, then I_{OFF} increases exponentially, see **Figure 2(a)**.

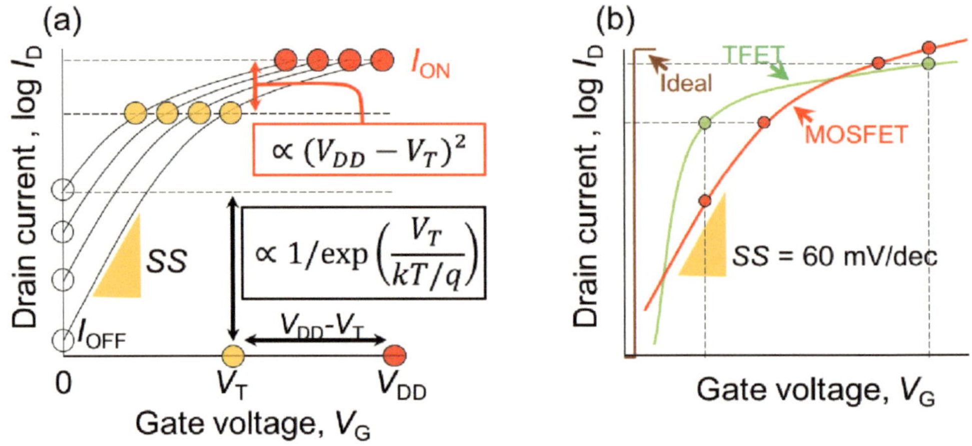

Figure 2.
(a) Transfer characteristics (drain current I_D vs. gate voltage V_G) of a MOSFET switch showing an exponential increase in I_{OFF} because of the thermionic limit of the subthreshold swing SS. Here the simultaneous scaling down of the supply voltage V_{DD} and the threshold voltage V_T, keeping the same I_{ON} by keeping the (V_{DD}-V_T) constant. (b) Qualitative comparison of the MOSFET switch (red) with a TFET (green) which has a steep off-on transition and a lower I_{OFF}. At low V_G, because of the subthermionic SS, the TFET offers a better performance and power reduction. At high V_G, the MOSFET switch becomes a better solution for higher performance thanks to the higher I_{ON}.

A possible way of reducing the voltage supply without performance loss is to increase the turn-on steepness, which means decreasing the average SS below the SS_{min}. Such devices, called steep-slope switches, are expected to effectively enable power scaling. Because of these MOSFET limitations, other device architectures are under active investigation, including the negative-capacitance FET (NC-FET) and the tunnel FETs (TFETs) [3].

3. Tunnel field-effect transistors

In contrast to MOSFETs, where charges are thermally injected over a potential barrier, the primary injection mechanism is band-to-band tunneling (BTBT), i.e. charge carriers transfer from one energy band into another. This tunneling mechanism was first identified by Zener in 1934 [4].

A typical TFET is composed of a p-i-n structure with a gated intrinsic region, see **Figure 3(a)**. Its working mechanism can be explained as follows. When a low voltage is applied to the gate, electrons tunneling from the valence band of the source to the conduction band of the drain is suppressed due to the gap in the intrinsic region, see **Figure 3(b)**. This is the off-state. When the potential applied to the gate brings the conduction band of the intrinsic limit at the same level as the source valence band, electrons can easily tunnel from source to drain, see **Figure 3(b)**. This is the on-state. In the ideal case, the transition from the off-state to the on-state is very fast, since the thermal tail of the injected electrons is cut by the top of the valence band in the source and the off-current is exponentially suppressed when the source Fermi level is within the gap of the intrinsic region. This would allow, in principle, very low SS, below SS_{min}, see **Figure 2(b)**.

Here, we briefly summarize the history of TFETs. The gated p-i-n structure was proposed in 1978 by Quinn et al. [5]. In 1992, Baba [6] fabricated TFETs called surface tunnel transistors in group III-V materials. In 1995, Reddick and Amaratunga [7] reported experiments on silicon surface tunnel transistors. In 2000, Hansch et al. [8] published experimental results on a reverse-biased vertical silicon

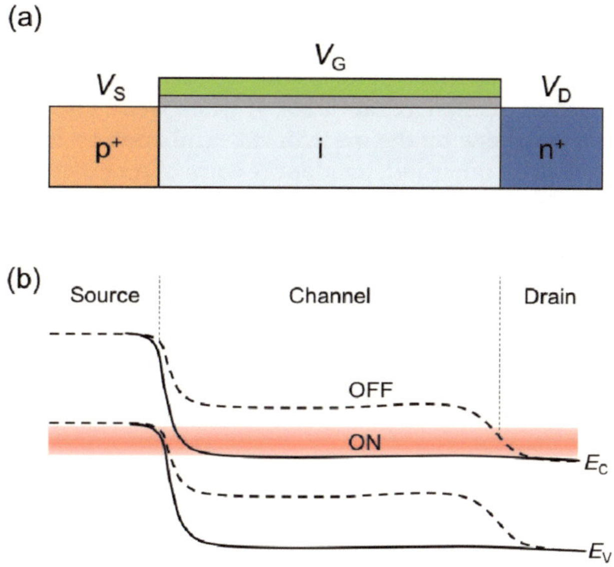

Figure 3.
(a) Schematic cross-section of n-type TFET with applied source (V_S), gate (V_G) and drain (V_D) voltages. (b) Schematic energy band profile for the off-state (dashed lines) and the on-state (solid lines) in a n-type TFET.

TFET fabricated by molecular beam epitaxy. Aydin et al. [9] fabricated lateral TFETs on silicon-on-insulator in 2004. Recently, TFETs fabricated in various materials (carbon, silicon, SiGe and group III-V materials) have emerged experimentally as the most promising candidates for switches with ultralow standby power and sub-0.5 V supply voltage.

The goals for TFET optimization are to simultaneously achieve the highest possible I_{ON}, the lowest average SS over many orders of magnitude of drain current, and the lowest possible I_{OFF}. For TFETs, SS decreases with the gate voltage, therefore they are naturally optimized for low-voltage operation. To achieve a high tunneling current and a steep slope, the transmission probability of the tunneling barrier should pass from 0 to close to 1 for a small change in gate voltage around the threshold potential. This requires a strong modulation of the channel bands by the gate and a very thin channel barrier.

As mentioned above, there have already been many experimental attempts to build TFETs with bulk silicon and III-V group materials. Even though encouraging experimental results have been reported for the on-current and SS in Si- and III–V-based TFETs, these devices are very demanding in terms of gate control [10]. Moreover, their transfer characteristics can be seriously degraded by the presence of interface or bulk defects enabling inelastic trap-assisted tunneling in the OFF-state [11, 12].

The 2-D materials (2DMs) may overcome some of the above issues [13], and have great potential for TFETs, due to their scalability and absence of dangling bonds at interface. They can be stacked to form a new class of tunneling transistors based on an interlayer tunneling occurring in the direction normal to the plane of the 2DMs [14]. In the next Section, we will review the properties of various different 2DMs and their applications.

4. 2-D materials

For a long time, the 2-D materials were thought to be unstable [15]. In the first half of the last century, scientists predicted [16] that a 2DM would likely disintegrate at finite temperature under the displacement of lattice atoms caused by thermal fluctuations. This theory was further supported by experiments observing that the melting temperature of thin film materials rapidly reduce with decreasing film thickness.

This belief remained unchanged until 2004, when Geim and Novoselov successully isolated graphene by the mechanical exfoliation technique [17, 18]. Although there have been other independent reports of monolayer carbon materials isolation [19, 20], some even long before the reports from the Manchester group, the works in 2004 and 2005 unveiled the unusual electronic properties of graphene, thus generating an intensive research by physicists and chemists in the field of 2DMs and inspiring 2DMs-based nanoelectronics [21]. Since then, we have seen an exponential increase in the research activity in graphene and other 2DMs, such as the transition metal dichalcogenides (TMDs), hexagonal boron nitride (*h*-BN), black phosphorus, silicene and gemanene [22–25].

4.1 Electronic properties of graphene

In 1946, Wallace first calculated the band structure of graphene and showed uncommon semimetallic behavior in this, at that time imaginary, 2-D material [26]. Actually, Wallace's studies of graphene served him as a starting point to study graphite. Between 1957 and 1958, other works by McClure [27] and Slonczewski and Weiss [28] followed.

One of the most interesting aspects of the graphene is its low-energy band structure [21, 29], which is linear around the K and K′ points of the Brillouin zone. In neutral graphene, the Fermi energy crosses exactly these points, which are also called Dirac points, because there electrons can be described by a 2-D Dirac Hamiltonian for masseless fermions, except for the fact that in graphene the electrons move with a speed v_F, which is 300 times smaller than the speed of light, and that the spin degree of freedom in replaced by the so-called pseudospin degree of freedom corresponding to the two graphene sublattices. This determines the observation of very unconventional properties with respect to the usual 2-D electron gases (2DEGs) obtained in doped semiconductor heterostructures. For example, in the presence of a strong magnetic field, Landau levels form at both positive and negative energies (with respect to the Dirac points) at energies proportional to the square root of the magnetic field and of the level index. This gives rise to the anomalous integer quantum Hall effect [30, 31], which, compared to the case of 2DEGs, can be observed at relatively low magnetic field and high temperature, with interesting perspective applications in metrology [32]. The linear dispersion of the graphene energy bands also entails a very high electron mobility, up to 200,000 $cm^2/(Vs)$ at low temperature for suspended graphene [33].

Another interesting property of graphene is when laterally confined into nanoribbons, its electronic and transport properties are strongly affected by the geometry of the edges (armchair, zigzag or mixed) and the nature of their passivation. For example, under certain condition the ribbon can show a band gap, whose size is proportional to the inverse of the ribbon width. Such a gap might be important for applications in logic devices [34], which are however compromised by the huge mobility degradation due to the increase of the effective mass from one side, and the presence of edge roughness from the other.

4.2 Electronic properties of 2-D TMDs

Unlike traditional bulk semiconductors, such as Si and III-V group materials, 2-D TMDs have ultra thin thickness, no surface dangling bonds, and high flexibility, which make them promising candidates [35] to solve the new challenges the semiconductor industry is facing today, including short-channel effects, power dissipation, integration and flexible applications. Recently, one of the large research interests in the field of 2DMs is the understanding of the fundamental electronics properties.

MoS_2 is a typical and well-studied TMD material [36, 37]. Its layers consist of hexagons with the Mo and S atoms located at alternating corners. The most striking feature of bulk MoS_2 is that, compared to zero-bandgap graphene and insulating *h*-BN, it is a semiconductor with an indirect band gap of 1.29 eV [38]. Several studies have confirmed a transition from an indirect band gap to a direct band gap for MoS_2 as the thickness of bulk MoS_2 is decreased to that of a monolayer. Such a similar transition is also demonstrated for other TMD materials. Kuc et al. [39] performed an extended study of the influence of quantum confinement on the electronic structures of monolayer and few-layer MS_2 (M = W, Nb, Re) using first-principles calculations. They found that WS_2, which is similar to MoS_2, exhibits an indirect (bulk, E_g =1.3 eV) to direct (monolayer, E_g =2.1 eV) band gap transition. The band alignment of various monolayer semiconducting TMDs and monolayer SnS_2 has been obtained from first-principle calculations [40].

Electrical characterizations of single-layer MoS_2 have shown n-type conductivity with a room temperature mobility in the range of 0.5–3 $cm^2/(V{\cdot}s)$ [18, 41]. Using a halfnium oxide gate dielectric, Radisavljevic et al. [42] demonstrated a room-temperature mobility of single-layer MoS_2 is at least 200 $cm^2/(V{\cdot}s)$, similar to that of graphene nanoribbons, but still much lower than that of either pristine graphene or Si transistors. Because of the low mobility, MoS_2 transistors are probably more suited for low-power applications compared to Si transistors, rather than for high-performance applications.

4.3 2DM-based MOSFET

Apart from the main technological challenges in geometric scaling, a bigger intrinsic challenge is represented by the material properties: carrier mobility in silicon strongly decreases with body thickness reduction or increased doping, thus undermining possible improvements in the device switching speed. In this context, the 2DMs with their extreme thinness can serve as a convenient alternative. Considering digital electronic applications, graphene-based FETs cannot conform to the ITRS requirements because of its zero band gap, which leads to at most a few tens I_{ON}/I_{OFF} ratio, and large I_{OFF}. Many attempts have been made in order to open an energy gap in graphene, for example by applying a strong electric field over bilayer graphene [43], by quantum confinement in graphene nanoribbons with well-controlled width [44–46], by doping graphene with adatoms like boron atoms [46, 47].

The advantage of 2-D TMDs over graphene is the existence of an energy gap, which is crucial for low I_{OFF} and high I_{ON}/I_{OFF} ratio. Comparing to bulk materials, semiconducting 2-D TMDs have unique features that make them attractive as a channel material for FETs: their atomic thinness, the lack of dangling bonds, and a mobility comparable to Si [48]. One of the earliest uses of TMDs in FETs was reported in 2004, where WSe_2 crystals showed mobility comparable to the best single-crystal Si FETs (up to 500 $cm^2/(Vs)$ for p-type conductivity at room temperature), and a 10^4 I_{ON}/I_{OFF} ratio at a temperature of 60 K [49]. This result was soon followed by devices based on thin films of MoS_2 with a back-gated configuration, resulting in mobility values in the range 0.1–10 $cm^2/(Vs)$ [18, 50].

The first implementation of a top-gated transistor based on monolayer MoS_2 was reported by Kis et al. [51]. This device showed excellent I_{ON}/I_{OFF} ratio (10^8), n-type conduction, room-temperature mobility of >200 $cm^2/(Vs)$ and SS of 74 mV/dec [51]. The top-gated geometry allowed a reduction in the voltage necessary to switch the device and the integration of multiple devices on the same substrate. The high-*k* dielectric used in this device, HfO_2, also gave the additional benefit of improving the mobility of monolayer MoS_2. Top-gating with a high-*k* dielectric was also used in a p-type FET with an active channel made of a monolayer flake of WSe_2, which showed room-temperature hole mobility of 250 $cm^2/(Vs)$, close to 60 mV/dec SS and 10^6 I_{ON}/I_{OFF} ratio [52].

Although 2-D TMDs may not compete with conventional III–V transistors on the mobility values, for devices with very short channel length the transport becomes nearly ballistic, thus mitigating this issue. The ultimate thin body of 2-D TMDs provides high degree of electrostatic control that is important for device scaling and for reducing the SCEs. Furthermore, the relatively large effective mass for electrons and holes in TMDs represents an advantage, since a larger effective mass implies a larger density of states and therefore a larger ballistic I_{ON}. Taking into account the above facts, 2-D TMDs are promising candidates for future digital electronics. Recently, a demonstration of extremely scaled transistor based on a MoS_2 channel and 1-nanometer carbon nanotube gate was successfully implemented by Desai et al. [53]. This device exhibited SS of 65 mV/dec (near ideal value for MOSFET), and I_{ON}/I_{OFF} ratio of 10^6.

5. TFETs based on 2-D material vdWHs

One promising alternative energy-efficient switch to a MOSFET is TFET. To date, significant progress has been made in TFETs built on bulk Si, Ge, and III-V compound materials. TFETs built on bulk Si usually suffer from poor on-state current because of its large and indirect bandgap. To enhance tunneling, heterostructures have been adopted using III-V compound materials. The on state current of the TFET built on InAs/InGaAsSb/GaSb heterostructure nanowire has reached several μA/μm [54]. Despite tremendous advancements in the field, achieving simultaneously high on state current density and sub-thermionic SS over multiple current decades remains an open challenge for the TFETs. Various studies have demonstrated that the band edge roughness of the semiconductor in combination with the trap states at the surface and interfaces are limiting SS of the TFETs [55]. In addition, doping is known to further reduce band edge sharpness given the random distribution of the dopant atoms in the lattice. To overcome these fundamental limitations, 2-D layered materials show promise toward obtaining steep band edge tunneling devices, since they intrinsically have atomic level flatness.

5.1 Vertical vdWH TFET

A promising research direction is the stacking of different 2-D layered materials and/or 3-D bulk materials for fabricating vertical heterostructures with novel operation principles [56, 57]. A TFET was demonstrated to work by Britnell et al. [58] using two independently controlled graphene layers separated by thin hexagonal boron nitride ($h-$ BN) layers acting as tunneling barrier. Recently, a vertical TFET based on bilayer MoS_2 as the channel and degenerately doped p-type Ge as the source, was fabricated by Sarkar et al. [59] and showed rather low SS of 31 mV/dec over 4 decades, and quite high on-current in TFETs, with V_{DD} as low as 0.1 V. The van der Waals gap between the 2-D MoS_2 and Ge acts as an extremely thin tunneling barrier, which enhances the tunneling and the on-current.

Early preparations of vdWH required the layer-by-layer exfoliation and restacking. However, restacking might bring adsorbates between layers, detrimental for creating vdWHs with atomically sharp interfaces, demonstrated by Haigh et al. [60]. The intercalation of selected 2D materials with alkali metal ions offered an alternative way to vdWH construction. To boost on state current, the materials with high carrier mobilities such as graphene and Ge are suitable for the injection layers, while the semiconductors and insulators such as hBN and MoS_2 tend to be chosen as barrier layers. The carriers mainly transport between the two 2-D layers via interlayer tunneling. Yu et al. have demonstrated vdWH FET using a graphene/MoS_2 junction. However, the working mechanism of this device was proven to be the thermionic emission acrossing a Schottky barrier [61]. Similar tunneling phenomenon had also been observed in a WSe_2-based vertical graphene-TMD vdWH transistor by Shim et al. [62]. An extraordinarily large I_{ON}/I_{OFF} ratio of 5×10^7 was achieved at 180 K. The negative differential resistance (NDR) device made of BP/ReS_2 vdWH was reported to show rather high peak-to-valley current ratio values of 4.2 and 6.9 at room and low temperatures, showing advantages for future multi-valued logic devices [63]. Vertical vdWH TFET can be constructed using different 2-D semiconductor layers forming type-II (staggered) heterostructure band alignment. Under gate voltages, the band alignment passes from type-II to type-III (broken gap), such transition ignites the interlayer tunneling. Yan et al. reported a practical vertical n-type vdWH TFET built with $SnSe_2$/WSe_2 vdWH, showing a minimum SS of 37 mV/dec and a I_{ON}/I_{OFF} ratio exceeding 1×10^6 [64].

Similar with TFETs, there are recently some other different types of vdWH transistors that also operate with the assistance of tunneling. Here, we introduce two of them namely Dirac-source and cold source FETs. Quite recently, Peng's group proposed a novel Dirac-source FET which can operate below the thermionic limit for several decades at room temperature [65]. In the source material, electrons follow the thermal Boltzmann statistics, leaving an exponential tail inside the conduction/valence band. This tail is exactly the origin of the thermionic limit of SS = 60 mV/dec at room temperature. Here, the idea of the Dirac-source is to make this tail decay faster than the exponential function, by exploiting the particular energy dependence of DOS of graphene. The DOS decreases rapidly with energy approaching the Dirac point. For a n-type FET, if we set the Fermi level in graphene below the Dirac point, we can press down the thermal tail and decrease SS to below 60 mV/dec at room temperature. Extending on this concept, the cold source FETs have been proposed and investigated by Logoteta et al. [66], employing MoS_2-nanoribbon as the source material, which exploits a narrow-energy conduction band to intrinsically filter out the thermionic tail of the electron energy distribution. It should be noted that in principle, the Dirac-source and cold source FETs do not purely work on the tunneling mechanism to achieve sub-thermionic SS. Rather, the switching is obtained by modulating the height of a potential barrier under the gate electrode, exactly as in MOSFETs. The authors of these works expect that the operation of these devices to be significantly less sensitive to the performance-degrading factors plaguing the TFETs, such as the traps, band tails and roughness.

6. Quantum transport modeling

To solve the emerging problems and to investigate improved strategies both need advanced predictive simulation and modeling as theoretical guidance. The accuracy of the theory model will influence the exploring directions to a great extent. To properly describe and model the tunneling current flow in TFETs, we need to develop a simulation approach able to take into account quantum phenomena as well as non-ideality effects due to phonon assisted tunneling. With appropriate simplifications to overcome the computational difficulties, the Non-Equilibrium Green's Function (NEGF) formalism provides a suitable framework to simultaneously treat the quantum transport of coherent carriers and the impact of diffusive phenomena such as electron-phonon interaction. The concept and the first applications of the NEGF were given by Schwinger [67], Kadanoff and Baym [68], Fujita [69], and Keldysh [70] at the beginning of the 1960s. The main advantages of NEGF are that it is full quantum, adaptable to different Hamiltonian types (effective mass, $k \cdot p$, tight-binding), and able to deal with many-body interactions through the introduction of self-energy operators [71–74].

Let us start by considering the Hamiltonian operator $\hat{H} = \hat{H}_0 + \hat{V}$. To evaluate the time evolution of the Green's function $G(\mathbf{r}, t; \mathbf{r}', t')$ in the Heisenberg picture, the time derivative of $G(\mathbf{r}, t; \mathbf{r}', t')$ can be obtained

$$\begin{aligned}\left(\mathrm{i}\hbar\frac{\mathrm{d}}{\mathrm{d}t} - H_0(\mathbf{r})\right)G(\mathbf{r}, t; \mathbf{r}', t') &= \delta(t - t')\delta(\mathbf{r} - \mathbf{r}') \\ &\quad -\mathrm{i}\int \mathrm{d}\mathbf{r}_1 \int_C \mathrm{d}t_1 V(\mathbf{r} - \mathbf{r}_1)\delta(t_1 - t)G^{(2)}(\mathbf{r}_1, t_1, \mathbf{r}, t; \mathbf{r}_1, t_1, \mathbf{r}', t'),\end{aligned} \tag{4}$$

where $G^{(2)}$ is the *two-particle* Green's function. By further differentiating $G^{(2)}$, we get an equation of motion containing the three-particle Green's function, whose equation of motion depends on the four-particle Green's function, and so on.

Instead of solving the infinite hierarchy of the Green's functions, the irreducible self-energy is introduced, which is represented with the symbol Σ, and which is a functional of the single-particle Green's function G. We can replace the right-hand side expression $-\mathrm{i}V(\mathbf{r} - \mathbf{r}_1)\delta(t_1 - t)G^{(2)}(\mathbf{r}_1, t, \mathbf{r}, t; \mathbf{r}_1, t, \mathbf{r}', t')$ by $\Sigma(\mathbf{r}, t, \mathbf{r}_1, t_1)G(\mathbf{r}_1, t_1; \mathbf{r}', t')$ into (4) to get

$$\begin{aligned}\left(\mathrm{i}\hbar\frac{\mathrm{d}}{\mathrm{d}t} - H_0(\mathbf{r})\right)G(\mathbf{r}, t; \mathbf{r}', t') &= \delta(t - t')\delta(\mathbf{r} - \mathbf{r}') \\ &\quad - \int \mathrm{d}\mathbf{r}_1 \int_C \mathrm{d}t_1 \Sigma(\mathbf{r}, t, \mathbf{r}_1, t_1)G(\mathbf{r}_1, t_1; \mathbf{r}', t').\end{aligned} \tag{5}$$

The Green's function G is defined by the contour ordering along the contour $C = C_+ + C_-$ close to the real axis in the complex time plan. Since it is not obvious to keep track of the time-branch in the evaluation of the integral, four new Green's functions are defined: G_c the chronologically time-ordered Green's function, G_a the anti-chronologically time-ordered Green's function, $G^<$ the lesser Green's function and $G^>$ the greater Green's function. These four functions are not independent since $G_c + G_a = G^> + G^<$. The greater and lesser Green's functions are directly related to the hole density and electron density in the system. We also define the advanced and retarded Green's functions $G^A = G_c - G^>, G^R = G_c - G^<$, with $G^R - G^A = G^> - G^<$. We can define the same quantities for the Σ self-energy leading to the lesser $\Sigma^<$, the greater $\Sigma^>$, the advanced Σ^A and the retarded Σ^R self-energies. We also have the relation $\Sigma^R - \Sigma^A = \Sigma^> - \Sigma^<$.

Under steady-state condition, the Green's functions depend on time difference $\tau = t - t'$. We can Fourier transform the time difference coordinate τ to energy $G^{R,A,<,>}(\mathbf{r}, \mathbf{r}'; E) \equiv \int \frac{\mathrm{d}\tau}{\hbar} e^{\mathrm{i}E\tau/\hbar} G^{R,A,<,>}(\mathbf{r}, \mathbf{r}'; \tau)$, as well as for the self-energies $\Sigma^{R,A,<,>}(\mathbf{r}, \mathbf{r}'; E) \equiv \int \frac{\mathrm{d}\tau}{\hbar} e^{\mathrm{i}E\tau/\hbar} \Sigma^{R,A,<,>}(\mathbf{r}, \mathbf{r}'; \tau)$. Using Langreth's rules [75], the equations of motion for the Green's functions with respect to time t become

$$\begin{cases} (E - H_0(\mathbf{r}))G^R(\mathbf{r}, \mathbf{r}'; E) = \delta(\mathbf{r} - \mathbf{r}') - \int \mathbf{dr}_1 \Sigma^R(\mathbf{r}, \mathbf{r}_1; E) G^R(\mathbf{r}_1, \mathbf{r}'; E), \\ (E - H_0(\mathbf{r}))G^A(\mathbf{r}, \mathbf{r}'; E) = \delta(\mathbf{r} - \mathbf{r}') - \int \mathbf{dr}_1 \Sigma^A(\mathbf{r}, \mathbf{r}_1; E) G^A(\mathbf{r}_1, \mathbf{r}'; E), \\ (E - H_0(\mathbf{r}))G^<(\mathbf{r}, \mathbf{r}'; E) = \int \mathbf{dr}_1 \Sigma^R(\mathbf{r}, \mathbf{r}_1; E) G^<(\mathbf{r}_1, \mathbf{r}'; E) + \Sigma^<(\mathbf{r}, \mathbf{r}_1; E) G^A(\mathbf{r}_1, \mathbf{r}'; E), \\ (E - H_0(\mathbf{r}))G^>(\mathbf{r}, \mathbf{r}'; E) = \int \mathbf{dr}_1 \Sigma^A(\mathbf{r}, \mathbf{r}_1; E) G^>(\mathbf{r}_1, \mathbf{r}'; E) + \Sigma^>(\mathbf{r}, \mathbf{r}_1; E) G^R(\mathbf{r}_1, \mathbf{r}'; E). \end{cases} \tag{6}$$

The electron and hole concentration are respectively given by

$$\begin{cases} n(\mathbf{r}, t) = \langle \hat{\psi}^\dagger(\mathbf{r}, t) \hat{\psi}(\mathbf{r}, t) \rangle = -i\hbar G^<(\mathbf{r}, t; \mathbf{r}, t), \\ p(\mathbf{r}, t) = \langle \hat{\psi}(\mathbf{r}, t) \hat{\psi}^\dagger(\mathbf{r}, t) \rangle = +i\hbar G^>(\mathbf{r}, t; \mathbf{r}, t). \end{cases} \tag{7}$$

Under the steady-state condition, the relations can be expressed in the energy domain

$$\begin{cases} n(\mathbf{r}) = -\mathrm{i} \int \frac{\mathrm{d}E}{2\pi} G^<(\mathbf{r}, \mathbf{r}; E), \\ p(\mathbf{r}) = +\mathrm{i} \int \frac{\mathrm{d}E}{2\pi} G^>(\mathbf{r}, \mathbf{r}; E). \end{cases} \tag{8}$$

The total space charge density is therefore given by $\rho(\mathbf{r}) = e[p(\mathbf{r}) - n(\mathbf{r}) + N_D(\mathbf{r}) - N_A(\mathbf{r})]$, where e is the absolute value of the electron charge, N_D and N_A are the donors and acceptors concentrations.

The current density vector is given by

$$\vec{J} = \frac{e\hbar}{2m_0} \lim_{\mathbf{r}' \to \mathbf{r}} (\nabla_{\mathbf{r}'} - \nabla_{\mathbf{r}}) G^<(\mathbf{r}, t; \mathbf{r}', t). \tag{9}$$

In practice, to evaluate $\vec{J}$ the lesser Green's function is expanded in a local basis, and the ∇-operator and m_0 are replaced by appropriate matrix elements $h_{m,n}$.

Noticing that at steady-state the time derivative of the charge density is zero. This implies that the $\nabla_{\mathbf{r}} \cdot \vec{J} = 0$ and the stationary current should be conserved in the device. In the presence of elastic scattering alone, the current density is conserved for electron at any given energy. With the inelastic scattering, the total current (integrated over energy) should be conserved.

The G^R and G^A Green's functions define the spectral function as

$$A(\mathbf{r}, \mathbf{r}'; E) = \mathrm{i}\left[G^R(\mathbf{r}, \mathbf{r}'; E) - G^A(\mathbf{r}, \mathbf{r}'; E)\right]. \tag{10}$$

The spectral function provides information about the nature of the allowed electronic states, regardless of whether they are occupied or not, and can be considered as a generalized density of states. The diagonal elements of the spectral function give the local density of states (LDOS), $D(\mathbf{r}, E) = \frac{1}{2\pi} A(\mathbf{r}, \mathbf{r}; E)$. Therefore the trace of the spectral function gives the density of states $N(E) = \mathrm{Tr}[A(E)] = \int \mathbf{dr} A(\mathbf{r}, \mathbf{r}; E)$.

On the other hand, the charge density can be included into the Poisson equation to renew the electrostatic potential, as $\nabla_{\mathbf{r}} \cdot [\varepsilon(\mathbf{r}) \nabla_{\mathbf{r}} \varphi(\mathbf{r})] + \rho(\mathbf{r}) = 0$, with the charge density ρ, and the dielectric constant ε. Since the potential depends on the charge density, which is given by the $G^{<}$ Green's function, the exact Green's function G both determines and is determined by the potential ϕ. The coupling between Green's function and the Poisson's equation needs to be solved self-consistently.

The Hamiltonian in a tight binding (TB) form can be obtained from the Wannier bases naturally. This procedure is called the maximally localized Wannier functions (MLWFs) method [76]. Or it can be obtained by the effective-mass approximation (EMA) and the k·p methods. These methods have been successfully utilized to predict the quantum transport behaviors of the devices based on many traditional and 2-D materials [77–87]. It should be noted that in the first principles calculations periodic boundary condition is necessary. As mentioned above, the TFET performance can be improved significantly by vdWHs that are not suitable for the first principles calculations.

7. Conclusions

We have given a literature survey on recently developed TFETs based on 2-D materials and their vdWHs. Compared with conventional MOSFETs, TFETs mainly work through the BTBT mechanism, resulting in large I_{ON}/I_{OFF} ratios within small supply voltages. To boost the on- and off-state behaviors simultaneously, heterojunctions should be adopted in the TFET design. Then, various novel TFETs based on the vdWHs are studied from structures to working mechanisms. We have also presented the quantum transport simulation method based on the NEGF formalism. However, no 2-D materials vdWH TFET in the experiments exhibits a satisfactorily overall performance. There is still a long way to realize 2-D TFET application. However, we hold an optimistic attitude toward vdWH TFET.

Acknowledgements

J.C. acknowledges support in part from the Natural Science Foundation of Jiangsu Province under grant number BK20180456 and from the Key Laboratory for Information Science of Electromagnetic Waves (MoE) under grant numbers EMW201906.

Conflict of interest

The authors declare no conflict of interest.

Author details

Jiang Cao
School of Electronic and Optical Engineering, Nanjing University of Science and Technology, Nanjing, China

*Address all correspondence to: jiang.cao@njust.edu.cn

References

[1] Gargini P. ITRS Roadmap: 2013 Update. 2013. Available from: www.itrs.net

[2] Balestra F, Cristoloveanu S, Benachir M, Brini J, Elewa T. Double-gate silicon-on-insulator transistor with volume inversion: A new device with greatly enhanced performance. IEEE Electron Device Letters. 1987;**8**(9): 410-412

[3] Cristoloveanu S, Wan J, Zaslavsky A. A review of sharp-switching devices for ultra-low power applications. IEEE Journal of the Electron Devices Society. 2016;**4**(5):215-226

[4] Zener C. A theory of the electrical breakdown of solid dielectrics. Proceedings of the Royal Society A: Mathematical, Physical and Engineering Sciences. 1934;**145**(855):523-529

[5] Quinn JJ, Kawamoto G, McCombe BD. Subband spectroscopy by surface channel tunneling. Surface Science. 1978;**73**(C):190-196

[6] Baba T. Proposal for surface tunnel transistors. Japanese Journal of Applied Physics. 1992;**31**(4):L455-L457

[7] Reddick WM, Amaratunga GAJ. Silicon surface tunnel transistor. Applied Physics Letters. 1995;**67**(1995):494

[8] Hansch W, Fink C, Schulze J, Eisele I. Vertical MOS-gated Esaki tunneling transistor in silicon. Thin Solid Films. 2000;**369**(1):387-389

[9] Aydin C, Zaslavsky A, Luryi S, Cristoloveanu S, Mariolle D, Fraboulet D, et al. Lateral interband tunneling transistor in silicon-on-insulator. Applied Physics Letters. 2004;**84**(10):1780

[10] Esseni D, Pala MG, Rollo T. Essential physics of the OFF-state current in nanoscale MOSFETs and tunnel FETs. IEEE Transactions on Electron Devices. 2015;**62**(9):3084-3091

[11] Esseni D, Pala MG. Interface traps in InAs nanowire tunnel FETs and MOSFETs—Part II: Comparative analysis and trap-induced variability. IEEE Transactions on Electron Devices. 2013; **PP**(99):1

[12] Esseni D, Pala MG. Interface traps in InAs nanowire tunnel FETs and MOSFETs - part II: Comparative analysis and trap-induced variability. IEEE Transactions on Electron Devices. 2013;**60**(9):2802-2807

[13] Jena D. Tunneling transistors based on graphene and 2-D crystals. Proceedings of the IEEE. 2013;**101**(7): 1585-1602

[14] Li MO, Esseni D, Snider G, Jena D, Xing HG. Single particle transport in two-dimensional heterojunction interlayer tunneling field effect transistor. Journal of Applied Physics. 2014;**115**(7):074508

[15] Landau LD, Lifshitz EM. Statistical Physics: Volume 5. Elsevier; 1980

[16] Landau L. Zur Theorie der Phasenumwandlungen II. Physikalische Zeitschrift der Sowjetunion. 1937;**11**:26-35

[17] Novoselov KS, Geim AK, Morozov SV, Jiang D, Zhang Y, Dubonos SV, et al. Electric field effect in atomically thin carbon films. Science. 2004;**306**(5696):666-669

[18] Novoselov KS, Jiang D, Schedin F, Booth TJ, Khotkevich VV, Morozov SV, et al. Two-dimensional atomic crystals. Proceedings of the National Academy of Sciences of the United States of America. 2005;**102**(30):10451-10453

[19] Van Bommel AJ, Crombeen JE, Van Tooren A. LEED and Auger electron

observations of the SiC(0001) surface. Surface Science. 1975;**48**(2):463-472

[20] Oshima C, Itoh A, Rokuta E, Tanaka T, Yamashita K, Sakurai T. Hetero-epitaxial-double-atomic-layer system of monolayer graphene/ monolayer h-BN on Ni(111). Solid State Communications. 2000;**116**(1):37-40

[21] Castro Neto AH, Guinea F, Peres NMR, Novoselov KS, Geim AK. The electronic properties of graphene. Reviews of Modern Physics. 2009;**81**(1): 109-162

[22] Miro P, Audiffred M, Heine T. An atlas of two-dimensional materials. Chemical Society Reviews. 2014;**43**(18): 6537-6554

[23] Kara A, Enriquez H, Seitsonen AP, Lew Yan Voon LC, Vizzini S, Aufray B, Oughaddou H. A review on silicene - New candidate for electronics. Surface Science Reports. 2012;**67**(1):1-18

[24] Xu M, Liang T, Shi M, Chen H. Graphene-like two-dimensional materials. Chemical Reviews. 2013; **113**(5):3766-3798

[25] Butler SZ, Hollen SM, Cao L, Cui Y, Gupta JA, Gutirrez HR, et al. Progress, challenges, and opportunities in two-dimensional materials beyond graphene. ACS Nano. 2013;7(4):2898-2926

[26] Wallace PR. The band theory of graphite. Physical Review. 1947;**71**(9): 622-634

[27] McClure JW. Band structure of graphite and de Haas-van Alphen effect. Physical Review. 1957;**108**(3): 612-618

[28] Slonczewski JC, Weiss PR. Band structure of graphite. Physical Review. 1958;**109**(2):272-279

[29] Goerbig MO. Electronic properties of graphene in a strong magnetic field. Reviews of Modern Physics. 2011;**83**(4): 1193-1243

[30] Novoselov KS, Geim AK, Morozov SV, Jiang D, Katsnelson MI, Grigorieva IV, et al. Two-dimensional gas of massless Dirac fermions in graphene. Nature. 2005;**438**(7065): 197-200

[31] Zhang YB, Tan YW, Stormer HL, Kim P. Experimental observation of the quantum Hall effect and Berry's phase in graphene. Nature. 2005;**438**(7065): 201-204

[32] Ribeiro-Palau R, Lafont F, Brun-Picard J, Kazazis D, Michon A, Cheynis F, et al. Quantum Hall resistance standard in graphene devices under relaxed experimental conditions. Nature Nanotechnology. 2015;**10** (September):965-971

[33] Du X, Skachko I, Barker A, Andrei EY. Approaching ballistic transport in suspended graphene. Nature Nanotechnology. 2008;**3**(8): 491-495

[34] Schwierz F. Graphene transistors. Nature Publishing Group. 2010;**5**(7): 487-496

[35] Schwierz F, Granzner R, Pezoldt J. Two-dimensional materials and their prospects in transistor electronics. Nanoscale. 2015;7:8261-8283

[36] Subbaiah YPV, Saji KJ, Tiwari A. Atomically thin MoS2: A versatile nongraphene 2D material. Advanced Functional Materials. 2016;**26**(13): 2046-2069

[37] Wang QH, Kalantar-Zadeh K, Kis A, Coleman JN, Strano MS. Electronics and optoelectronics of two-dimensional transition metal dichalcogenides. Nature Nanotechnology. 2012;7(11):699-712

[38] Dolui K, Pemmaraju CD, Sanvito S. Electric field effects on armchair MoS2

nanoribbons. ACS Nano. 2012;**6**(6): 4823-4834

[39] Kuc A, Zibouche N, Heine T. Influence of quantum confinement on the electronic structure of the transition metal sulfide ts2. Physical Review B. 2011;**83**(24):245213

[40] Cheng G, Zhang H, Wang W, Colombo L, Wallace RM, Cho K. Band alignment of two-dimensional transition metal dichalcogenides: Application in tunnel field effect transistors. Applied Physics Letters. 2013;**103**(5):53513

[41] Ghatak S. Atindra Nath Pal, and Arindam Ghosh. Nature of electronic states in atomically thin MoS2 field-effect transistors. ACS Nano. 2011; **5**(10):7707-7712

[42] Radisavljevic B, Whitwick MB, Kis A. Integrated circuits and logic operations based on single-layer MoS2. ACS Nano. 2011;**5**(12):9934-9938

[43] Castro EV, Novoselov KS, Morozov SV, Peres NMR, Santos JMBLD, Nilsson J, et al. Biased bilayer graphene: Semiconductor with a gap tunable by the electric field effect. Physical Review Letters. 2007;**99**(21): 216802

[44] Grassi R, Gnudi A, Gnani E, Reggiani S, Baccarani G. Mode space approach for tight binding transport simulation in graphene nanoribbon FETs. IEEE Transactions on Nanotechnology. 2011;**10**(3):371-378

[45] Gryglas-Borysiewicz M, Jouault B, TworzydłO J, Lewińska S, Strupiński W, Baranowski JM. Transport properties of disordered graphene layers. Acta Physica Polonica A. 2009;**116**(5): 838-840

[46] Ruitao LV, Terrones M. Towards new graphene materials: Doped graphene sheets and nanoribbons. Materials Letters. 2012;**78**:209-218

[47] Marconcini P, Cresti A, Triozon F, Fiori G, Biel B, Niquet YM, et al. Atomistic boron-doped grapheme field-effect transistors: A route toward unipolar characteristics. ACS Nano. 2012;**6**(9):7942-7947

[48] Bissessur R, Heising J, Hirpo W, Kanatzidis M. Toward pillared layered metal sulfides. Intercalation of the chalcogenide clusters Co(6)Q(8)(PR (3))(6) (Q=S, Se, and Te and R=alkyl) into MoS2. Chemistry of Materials. 1996;**8**(2):318

[49] Podzorov V, Gershenson ME, Kloc C, Zeis R, Bucher E. High-mobility field-effect transistors based on transition metal dichalcogenides. Applied Physics Letters. 2004;**84**(17): 3301-3303

[50] Ayari A, Cobas E, Ogundadegbe O, Fuhrer MS. Realization and electrical characterization of ultrathin crystals of layered transition-metal dichalcogenides. Journal of Applied Physics. 2007;**101**(1):014507

[51] Radisavljevic B, Radenovic A, Brivio J, Giacometti V, Kis A, Radisavljevic B, et al. Single-layer MoS2 transistors. Nature Nanotechnology. 2011;**6**(3):147-150

[52] Fang H, Chuang S, Chang TC, Takei K, Takahashi T, Javey A. High-performance single layered WSe 2 p-FETs with chemically doped contacts. Nano Letters. 2012;**12**(7):3788-3792

[53] Desai SB, Madhvapathy SR, Sachid AB, Llinas JP, Wang Q, Ahn GH, et al. MoS2 transistors with 1-nanometer gate lengths. Science. 2016;**354**(6308): 99-102

[54] Memisevic E, Hellenbrand M, Lind E, Persson AR, Sant S, Schenk A, et al. Individual defects in inas/ingaassb/gasb nanowire tunnel field-effect transistors operating below 60 mv/decade. Nano Letters. 2017;**17**(7):4373-4380

[55] Agarwal S, Yablonovitch E. Band-edge steepness obtained from esaki/backward diode current–voltage characteristics. IEEE Transactions on Electron Devices. 2014;**61**(5):1488-1493

[56] Geim AK, Grigorieva IV. Van der Waals heterostructures. Nature. 2014; **499**(7459):419-425

[57] Lu SC, Mohamed M, Zhu WJ. Novel vertical hetero- and homo-junction tunnel field-effect transistors based on multi-layer 2D crystals. 2D Materials. 2016;**3**(1):6

[58] Britnell L, Gorbachev RV, Jalil R, Belle BD, Schedin F, Mishchenko A, et al. Field-effect tunneling transistor based on vertical graphene heterostructures. Science. 2012; **335**(6071):947-950

[59] Sarkar D, Xie X, Liu W, Cao W, Kang J, Gong Y, et al. A subthermionic tunnel field-effect transistor with an atomically thin channel. Nature. 2015; **526**(7571):91-95

[60] Haigh SJ, Gholinia A, Jalil R, Romani S, Britnell L, Elias DC, et al. Cross-sectional imaging of individual layers and buried interfaces of graphene-based heterostructures and superlattices. Nature Materials. 2012; **11**(9):764-767

[61] Yu WJ, Li Z, Zhou H, Chen Y, Wang Y, Huang Y, et al. Vertically stacked multi-heterostructures of layered materials for logic transistors and complementary inverters. Nature Materials. 2013;**12**(3):246-252

[62] Shim J, Kim HS, Shim YS, Kang D-H, Park H-Y, Lee J, et al. Extremely large gate modulation in vertical graphene/wse2 heterojunction barristor based on a novel transport mechanism. Advanced Materials. 2016;**28**(26):5293-5299

[63] Shim J, Seyong O, Kang D-H, Jo S-H, Ali MH, Choi W-Y, et al. Phosphorene/rhenium disulfide heterojunction-based negative differential resistance device for multi-valued logic. Nature Communications. 2016;**7**(1):13413

[64] Yan X, Liu C, Li C, Bao W, Ding S, Zhang DW, et al. Tunable snse2/wse2 heterostructure tunneling field effect transistor. Small. 2017;**13**(34):1701478

[65] Qiu C, Liu F, Xu L, Deng B, Xiao M, Si J, et al. Dirac-source field-effect transistors as energy-efficient, high-performance electronic switches. Science. 2018;**361**(6400):387-392

[66] Logoteta D, Pala MG, Choukroun J, Dollfus P, Iannaccone G. A steep-slope mos2-nanoribbon mosfet based on an intrinsic cold-contact effect. IEEE Electron Device Letters. 2019;**40**(9): 1550-1553

[67] Kumar D. Brownian motion of a quantum particle. Physical Review A. 1984;**29**(3):1571-1573

[68] Schieve WC, Horwitz LP. Quantum Statistical Mechanics. Benjamin; 2009

[69] Fujita S. Partial self-energy parts of Kadanoff-Baym. Physica. 1964;**30**(4): 848-856

[70] Keldysh LV. Diagram Technique for Nonequilibrium Processes. Soviet Physics - JETP. 1964;**47**(4):1515-1527

[71] Economou EN, Cohen MH. Existence of mobility edges in Anderson's model for random lattices. Physical Review B. 1972;**5**(8):2931-2948

[72] Mahan GD. Many-Particle Physics. 3rd ed. Springer Science & Business Media; 2000

[73] DiVentra M. Electrical Transport in Nanoscale Systems, Volume 14. Cambridge University Press Cambridge; 2008

[74] Stefanucci G, van Leeuwen R. Nonequilibrium Many-Body Theory of Quantum Systems: A Modern Introduction, Volume 54. Cambridge University Press; 2013

[75] Haug H, Jauho A-P, Cardona M. Quantum kinetics in transport and optics of semiconductors, Volume 2. Springer; 2008

[76] Marzari N, Mostofi AA, Yates JR, Souza I, Vanderbilt D. Maximally localized wannier functions: Theory and applications. Reviews of Modern Physics. 2012;**84**(4):1419

[77] Cao J, Logoteta D, Ozkaya S, Biel B, Cresti A, Pala M, et al. A computational study of van der Waals tunnel transistors: Fundamental aspects and design challenges. In: IEDM, International Electron Devices Meeting. IEEE; 2015. pp. 313-316

[78] Cao J, Logoteta D, Ozkaya S, Biel B, Cresti A, Pala MG, et al. Operation and design of van der Waals tunnel transistors: A 3-D quantum transport study. IEEE Transactions on Electron Devices. 2016;**63**(11):4388-4394

[79] Cao J, Cresti A, Esseni D, Pala M. Quantum simulation of a heterojunction vertical tunnel FET based on 2D transition metal dichalcogenides. Solid-State Electronics. 2016;**116**:1-7

[80] Cao J, Park J, Triozon F, Pala MG, Cresti A. Simulation of 2d material-based tunnel field-effect transistors: Planar vs. vertical architectures; 2018. p. 4. Available from: https://www.openscience.fr/IMG/pdf/iste_compona no18v1n

[81] Szabo A, Rhyner R, Luisier M. Ab-initio simulations of MoS2 transistors: From mobility calculation to device performance evaluation. In: Technical Digest - International Electron Devices Meeting, IEDM, Volume 2015. IEEE; 2015. pp. 30.4.1-30.4.4

[82] Szabo A, Koester SJ, Luisier M. Ab-initio simulation of van der Waals MoTe2-SnS2 heterotunneling FETs for low-power electronics. IEEE Electron Device Letters. 2015;**36**(5):514-516

[83] Cauley S, Luisier M, Balakrishnan V, Klimeck G, Koh CK. Distributed non-equilibrium Green's function algorithms for the simulation of nanoelectronic devices with scattering. Journal of Applied Physics. 2011;**110**(4):043713

[84] Luisier M, Klimeck G. Atomistic full-band simulations of silicon nanowire transistors: Effects of electron-phonon scattering. Physical Review B: Condensed Matter and Materials Physics. 2009;**80**(15):155430

[85] Luisier M, Klimeck G. Simulation of nanowire tunneling transistors: From the Wentzel-Kramers-Brillouin approximation to full-band phonon-assisted tunneling. Journal of Applied Physics. 2010;**107**(8):84507

[86] Fonseca JE, Kubis T, Povolotskyi M, Novakovic B, Ajoy A, Hegde G, et al. Efficient and realistic device modeling from atomic detail to the nanoscale. Journal of Computational Electronics. 2013;**12**(4):592-600

[87] Chen FW, Ilatikhameneh H, Ameen TA, Klimeck G, Rahman R. Thickness engineered tunnel field-effect transistors based on phosphorene. IEEE Electron Device Letters. 2016;**3106**(3): 130-133

Section 4

Organic Field Effect Transistors

Chapter 4

Crystal Polymorph Control for High-Performance Organic Field-Effect Transistors

Zhi-Ping Fan and Hao-Li Zhang

Abstract

Organic molecules are assembled together by weak non-covalent intermolecular interactions in solid state. Multiple crystalline packing states (crystal polymorphism) have commonly existed in the active layer for organic field-effect transistors (OFETs). Different polymorphs, even with the slightest changes in their molecular packing, can differ the charge transport mobility by orders of magnitude. Therefore, accessing new polymorphs can serve as a novel design strategy for attaining high device performance. Here, we review the state of the art in this emerging field of crystal polymorph control. We firstly introduce the role of polymorphism and the methods of polymorph control in organic semiconductors. Then we review the latest studies on the performance of polymorphs in OFET devices. Finally, we discuss the advantages and challenges for polymorphism as a platform for the study of the relationship between molecular packing and charge transport.

Keywords: organic field-effect transistors, organic semiconductors, polymorphism, structure-property relationship, carrier mobility

1. Introduction

An organic field-effect transistor (OFET) is a transistor using an organic semiconducting thin film as the active layer in its channel [1, 2]. Charge carriers are transported in the OFET active layer under the electric field. Through the design of new materials and the improvement of fabrication processes, many impressive developments in the field of OFETs have been achieved [3–6]. It has long been realized that the morphology of the active layer has a crucial impact on its charge transport properties. Tremendous efforts have been devoted to fabricate highly ordered crystalline films to achieve high device performance, including introduction of self-assembled monolayers [7–9], annealing [10, 11], off-center spin coating [4, 12, 13], and solution shearing [14, 15]. However, the lack of knowledge on the intrinsic properties of organic semiconductors remains the barrier for high-performance materials being efficiently developed.

Polymorphism of organic semiconductors has recently received much attention in the field of OFETs [16–18]. Different polymorphic crystals have the same molecular structure but a different molecular arrangement, which can be used as an ideal platform to correlate charge transport with respect to molecular arrangement. Through investigating OFETs with different polymorphs, the relationship between molecular packing and charge transport can be obtained. Recently, some ultra-high-mobility

OFETs have been obtained by controlling the polymorphic structures of organic semiconductors, revealing that the crystal polymorph control has become an efficient strategy for the manufacture of high-performance OFETs [4, 17].

1.1 What is polymorphism?

Polymorphism refers to the ability for the same compound to adopt multiple crystalline packing states. Organic molecules assemble into crystals by weak intermolecular interactions, typically via van der Waals and electrostatic interactions. Many thermodynamic and kinetics factors (such as temperature, solvent mixtures, speed of crystallization, seeding, and pH) can have significant impacts on crystal growth, leading to polymorphism prevalent among organic materials. For instance, a continuous investigation on polymorphism has indicated that approximately one-third of organic substances show polymorphism under normal pressure conditions [19, 20]. Different polymorphs often have distinct physical properties such as solubility, melting point, and electrical, optical, and mechanical properties [21]. The interest in polymorphism has increased significantly in recent years, particularly in the pharmaceutical and material science fields [17, 22].

1.2 The role of polymorphism in OFET

The charge transport property of organic semiconductors is sensitive to the molecular packing, where a slight change in molecular packing may result in huge difference in charge carrier mobility [23]. The side-chain engineering, which is efficient in tailoring the molecular packing, has been extensively applied to develop high-performance organic semiconductors [24–27]. However, the introduction of side chains alters the molecular structure, which makes the investigation on relationship between molecular packing and charge transport very complex.

Polymorphism offers an opportunity to tailor the molecular packing of a material, without affecting its chemical components. For example, rubrene can crystallize into three crystalline polymorphs, including an orthorhombic, a triclinic, and a monoclinic phase (**Figure 1a**) [28–30]. Taking advantage of polymorphism, it is possible to fabricate OFETs from the same organic semiconductor but with different polymorphs, hence, with different properties (**Figure 1b**). Importantly, by

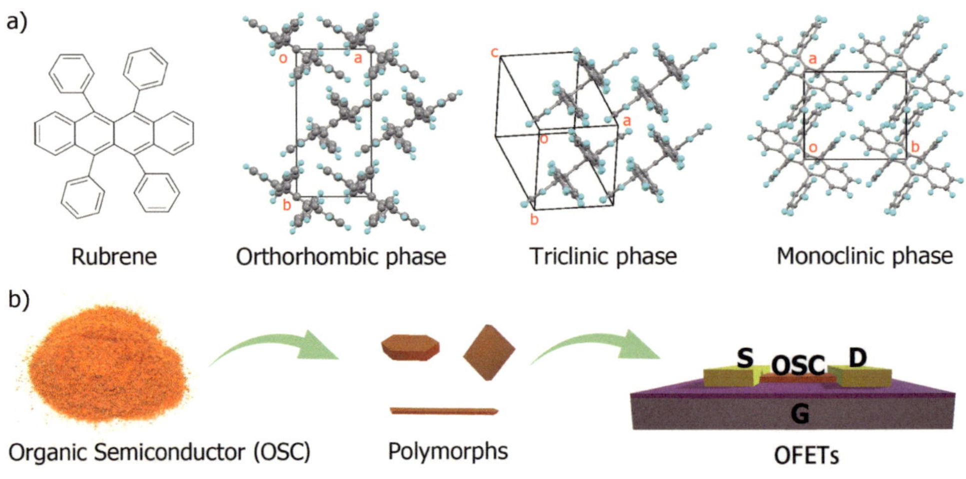

Figure 1.
(a) Molecular structure and the crystal phases of rubrene. (b) Schematic diagram for the fabrication of the OFETs from microcrystals of different polymorphs.

measuring charge transport performance in OFETs from different polymorphs, a direct relationship between molecular packing and charge transport can be established. Thus, many investigations on charge transport in different polymorphs have been performed in thin-film transistors, including some benchmark organic semiconductors like pentacene derivatives [15, 31, 32], rubrene [28–30], sexithiophene (6T) [33–36], and [1]benzothieno[3,2-b][1]benzothiophene (BTBT) derivatives [10, 37, 38]. However, the different factors affecting the device performance, including crystallinity, grain size, and grain boundaries, are difficult to be eliminated in thin-film transistors. Thus, tremendous efforts have been paid to manufacture OFETs from highly crystalline films even single crystals [39–44]. Especially, single-crystal OFETs from different polymorphs have attracted increasing attentions [45]. Compared to polycrystalline films, single crystals have the advantages of high molecular ordering and minimal grain boundaries, and its structure is much easier to determine.

There is no doubt that high-performance OFETs can be obtained by tailoring molecular packing motifs in the active layer. Among the investigations on polymorphism of organic semiconductors, several high-mobility OFETs have been manufactured by polymorph control, such as the TIPS-pentacene [46], C8-BTBT [4], C6-DBTDT [47], and TiOPC [48]. Tailoring polymorphs has become an emerging design strategy for high-performance organic electronics.

2. Methods of polymorph control in organic semiconductors

Though polymorphism is observed on many organic semiconductors, the fabrication of each polymorph with high purity is very difficult. For instance, even for the extensively studied organic semiconductors like pentacene and BTBT derivatives, only part of their polymorphs have been useful to establish the correlation between the material molecular packing and its charge transport properties [17]. The difficulties for the investigations on polymorphism include the fabrication of pure polymorphs and the determination of their crystal structures, where the polymorph control is fundamental. Some of the polymorph control methods most commonly applied to organic semiconductors are discussed below.

2.1 Solvent control

Solution process is important for the fabrication of organic semiconductor devices, which has the advantages of low-cost and large-area fabrication. In solution processes, solvent-induced polymorphism has been frequently observed in organic semiconductors such as DB-TTF [49], DT-TTF [50], TIPS-pentacene [51], and so on. Consequently, the solvent of choice for solution-processed organic semiconductors has become a commonly practiced method for high-throughput polymorph screening. For example, the triethylsilylethynyl anthradithiophene (TES-ADT) films can crystallize into two polymorphs from different solvents [52–54]. The polymorph selectivity for solution processes mostly relates to the polarity of solvents, while the concentration can also induce polymorphism [47, 55]. For instance, the C6-DBTDT molecules can crystallize into the α-phase and β-phase crystals from high concentration and dilute chlorobenzene solutions, respectively [47]. At the molecular level, the specific interactions between semiconductor and solvent molecules in the solution can induce the nuclei formation in a particular polymorph and therefore result in polymorph selectivity as a function of the solvent or the concentration [56].

2.2 Temperature control

The thermodynamic polymorphic selection is usually observed from deposition of organic semiconductors by physical vapor transport (PVT) processes [57]. In a study by Stevens et al., TMS-DBC crystals were synthesized in a crystallization tube by PVT method, where two polymorphs grew at different temperature regions [58]. The red low-temperature (LT) polymorph was obtained in regions with tube temperature about 25–65°C, while the yellow high-temperature (HT) polymorph grew in regions with temperature around 130–170°C. The two polymorphs were found extremely stable, which did not interconvert to the other crystal structure with subsequent solvent or thermal treatments. Temperature-induced polymorphism was also observed from TiOPC crystals fabricated by PVT technology [48]. Sheetlike α-phase TiOPC crystals were obtained on the substrate at the temperature zone of about 210°C, while ribbonlike β-phase crystals were grown at the temperature zone of about 180°C. For many polymorphic materials, most of the polymorphs are temperature sensitive, where thermal-induced phase transition is allowed. Consequently, temperature-induced polymorph selectivity is becoming an important strategy to access and study different polymorphs.

2.3 Crystallization through kinetics control

Crystallization through kinetics control is a powerful method for accessing metastable polymorphs, especially in thin-film geometry where kinetic trapping and thin-film confinement work in synergy. In a study by Wedl et al. [59], thin films of dihexylterthiophene (DH3T) were fabricated by spin coating, dip coating, drop casting, and physical vapor deposition, i.e., with very different crystallization speed. Three polymorphs of DH3T were discovered from the experiment, which was noted as the α-phase, β-phase, and the metastable thin-film phase. The crystallization speed was found to be a key parameter to control the respective polymorphs present in the films. The metastable thin-film phase was obtained from deposition techniques with fast crystallization speeds, such as physical vapor deposition, spin coating, drop casting with fast solvent evaporation, and dip coating with high withdrawal velocity. In contrast, a mixture of two stable polymorphs was observed in films fabricated by both drop casting and dip coating with slow evaporation of solvent. Crystallization kinetics control was also applied in a study by Giri et al. [14], wherein the solution shearing method with a function of shearing speed was utilized to fabricate thin films of TIPS-pentacene. Through fast solvent evaporation and quick crystallization, metastable states were kinetically trapped, which were relaxed to more stable states with toluene vapor annealing.

2.4 Templating via heterogeneous nucleation

Organic molecules are assembled by weak intermolecular interactions into their crystalline form, which can be affected by the molecule-surface interaction and the molecule-molecule interaction from additives. In other words, substrates and additives can act as templates to alter the crystal structure of organic semiconductors.

The substrate can promote heterogeneous nucleation of a particular polymorph due to specific interface interactions. On the substrate-thin-film interface, substrate-induced polymorphs (SIPs) are usually observed in the first few molecular layers, whose molecular packing are different from that in the bulk of the film. For example, two SIPs were firstly observed in thin films of pentacene, including a thin-film phase with d-spacing of 15.4 Å and a single-layer phase with d-spacing of 16.1 Å [32, 60, 61]. In contrast, the single-crystal phase of pentacene exhibits the d-spacing of 14.1 Å [62–64]. SIPs were also observed in films of the 2,7-dioctyloxy-BTBT (C_8O-BTBT-OC_8) derivative [10]. The C_8O-BTBT-OC_8 molecules adopt a slipped π-π stacking in bulk crystal while exhibiting a herringbone packing motif in the thin-film phase. However, the SIP of C_8O-BTBT-OC_8 is a metastable form induced by the substrate, which is converted to the bulk form in 6 months or by chloroform vapor annealing. Although SIPs are commonly observed in thin films of organic semiconductors, studies on the formation of SIPs are limited [65].

Polymer additives have been demonstrated effective for polymorph control. In a study on polymorphism of TIPS-pentacene, conjugated polymer additives including poly(3-hexylthiophene) (P3HT) and region random TIPS-pentacene-bithiophene polymer (PnBT-RRa) were used to template the formation of a polymorph [66]. Two new polymorphs of TIPS-pentacene were obtained, including the phase II and phase III that were synthesized form TIPS-pentacene/PnBT-RRa blend and TIPS-pentacene/P3HT blend, respectively. Compared to the phase I obtained from pure TIPS-pentacene, the phase II exhibits very small changes in crystal structure, which is attributed to the structural similarity between TIPS-pentacene and PnBT-RRa as well as their strong intermolecular interactions. In contrast, the crystal structure of phase III exhibits a large difference compared to the phase I, due to the lack of structural similarity between TIPS-pentacene and P3HT. These results demonstrate the possibility of using polymer additives as templates for accessing various polymorph phases.

2.5 Postdeposition control

Postdeposition processing is a commonly used method to investigate phase transition, from which various polymorphs can be accessed. Solvent and thermal annealing is widely used for postdeposition treatments to increase crystallinity, to enlarge grain size, and in some cases to alter the molecular packing in films of organic semiconductors [67–69]. In a study by Hiszpanski et al. [70], three polymorphs of contorted hexabenzocoronene (c-HBC) have been obtained by the application of both thermal and solvent vapor annealing. The P21/c polymorph of c-HBC is obtained from the amorphous film by thermal annealing. In contrast, polymorph II is accessed from either the amorphous film or the P21/c polymorph by tetrahydrofuran vapor annealing. Subsequently, thermal annealing of polymorph II always yields polymorph II. From an investigation by Campione et al., α-tetrathiophene (α-4T)/LT single crystals were obtained by the floating-drop technique, and a crystal to crystal phase transition was observed at 191°C from α-4T/LT to α-4T/HT, obtaining large and thick α-4T/HT single crystals [71].

2.6 Other methods

In addition to the major methods summarized above, a variety of novel approaches have been developed to control polymorphs. A direct strategy to change the crystal phase is to apply pressure on crystals. For instance, pressure-induced phase transition was observed from rubrene and fullerene derivatives [72–76]. Ito et al. and Segara et al. reported mechanically induced phase transitions [77, 78]. The nucleation of a polymorph is often affected by many processing parameters. For example, Lee et al. successfully fabricated a metastable polymorph of quaterrylene diimide by flow-assisted crystallization [79]. He et al. synthesized a new polymorph of Cl2-NDI by vapor sublimation in air [18]. Even light can have effects on polymorph formation. As investigated by Pithan et al., two polymorphs of sexithiophene were obtained in dark and illumination environments [33].

3. Charge transport in OFETs with different polymorphs

Polymorphism is an important platform to study the charge transport mechanism in organic semiconductors because in polymorphs the crystal structure is the only variable, while the chemical structure remains identical [80]. By studying a polymorphic organic semiconductor, changes in charge transport can be directly associated with the differences in molecular packing. Many studies have attempted without success to reveal the relationship between charge transport and molecular packing, including experimental studies and quantum chemical calculations on different polymorphs.

3.1 Theoretical studies

Charge transport in inorganic semiconductors is well described by the band-like charge transport model. In contrast, to describe charge transport in organic semiconductors is much complex, where both electron–electron and electron–phonon interactions must be taken into account [81, 82]. A phonon is described as a particle-like quantized mode of vibrational energy, which arises from oscillating atoms within a crystal. In organic crystals, the molecule packing can be significantly disrupted by thermal functions due to the weak intermolecular interactions. Therefore, the charge transport behavior of organic semiconductors is temperature dependent [83]. It turns out that a charge-hopping model is commonly observed at near and above room temperatures [84], while a band-like transport model is typically observed in single crystals at lower temperatures [85].

The OFET devices mostly work at near and above room temperatures and follow a hopping transport mechanism. The hopping mobility can be deduced from the Marcus theory through Eq. (1) [17, 86]:

$$\mu_{\text{hop}} = \frac{ea^2t^2}{k_B T\hbar}\left[\frac{\pi}{2E_{\text{pol}}\,k_B T}\right]^{\frac{1}{2}}\exp\left(-\frac{E_{\text{pol}}}{2K_B T}\right) \tag{1}$$

where e is electron charge, a is the spacing between molecules, t is the charge transfer integral, k_B is the Boltzmann constant, T is temperature, $\hbar$ is Planck constant, and E_{pol} is the polaron binding energy. The polaron binding energy is related to the reorganization energy (λ_{reorg}) via $E_{\text{pol}} = \lambda_{\text{reorg}}/2$. The reorganization energy (λ_{reorg}) depicts both the intramolecular and intermolecular contributions to the change in the geometry of the molecules during charge transfer [87]. In organic crystals, two major parameters affect the charge carrier mobility, which are the transfer integral (t) and the reorganization energy (λ_{reorg}). The transfer integral and the reorganization energy can be quantitatively determined by quantum chemical calculations, and therefore, the hopping mobility can be estimated. In general, organic semiconductors with higher transfer integral and lower reorganization energy have higher charge carrier mobility.

Taking advantages of the quantum chemical calculations, the relationship between molecular packing and charge transport can be examined. For instance, Bredas et al. simulated the sexithiophene dimers to understand the effect of the molecular overlap on the transfer integral between adjacent molecules [88]. First, the HOMO and LUMO energy splittings were examined with a variation in the intermolecular distance between the conjugated planes. Next, by keeping the bottom molecule and the intermolecular distance between the sexithiophene (6T) dimers (4 Å) fixed, the effects of lateral molecular displacement along the conjugated plane on the energy splittings were examined by moving the top 6T molecule. The energy splitting is directly proportional to the charge transfer integral (t). The electronic splittings of HOMO and LUMO exhibit an exponential decay as the intermolecular distance is increased. In contrast, the HOMO/LUMO energy splittings show large oscillations and tend to abate as the lateral displacement along the conjugated plane is increased. These results indicate that charge transport in organic semiconductors is sensitive to molecular packing. Consequently, altering the molecular packing, i.e., tuning the polymorph structure of organic semiconductors, provides an opportunity to improve the performance of OFETs.

3.2 Thin-film transistors with different polymorphs

Organic thin-film transistors are easy to fabricate in large area by solution processing and therefore have been widely used in various electronic devices like electronic paper [89] and medical sensors [90, 91]. However, the semiconductor processing method and conditions can greatly affect the molecular packing motif and consequently can dramatically affect the device performance (see Section 2). To date, the knowledge on how the charge transport in semiconductor films depends on molecular packing motif is still very limited. In this section, the relationship between molecular packing and charge transport will be discussed, giving some recent investigations as examples.

3.2.1 Pentacene

Pentacene is a benchmark organic semiconductor synthesized in 1912 [92], which exhibits excellent charge transport performance in thin-film transistor [93]. To date, there are five different polymorphs known for pentacene [9, 31, 32, 94]. As shown in **Figure 2**, the five polymorphs of pentacene are classified by their molecular layer thickness (d-spacing), where four thin-film forms exhibit d-spacing of 14.1, 14.4, 15.0, and 15.4 Å [31], and a monolayer form shows d-spacing of 16.1 Å [60]. However, among the five polymorphs, complete structural data have only been determined for the 14.1 and 14.4 Å polymorphs, where the 14.1 Å polymorph shares a similar packing as the single crystals. The single crystal of pentacene was reported with mobility around 5–40 $cm^2\ V^{-1}\ s^{-1}$ [43, 95]. A recent study by Ji et al.

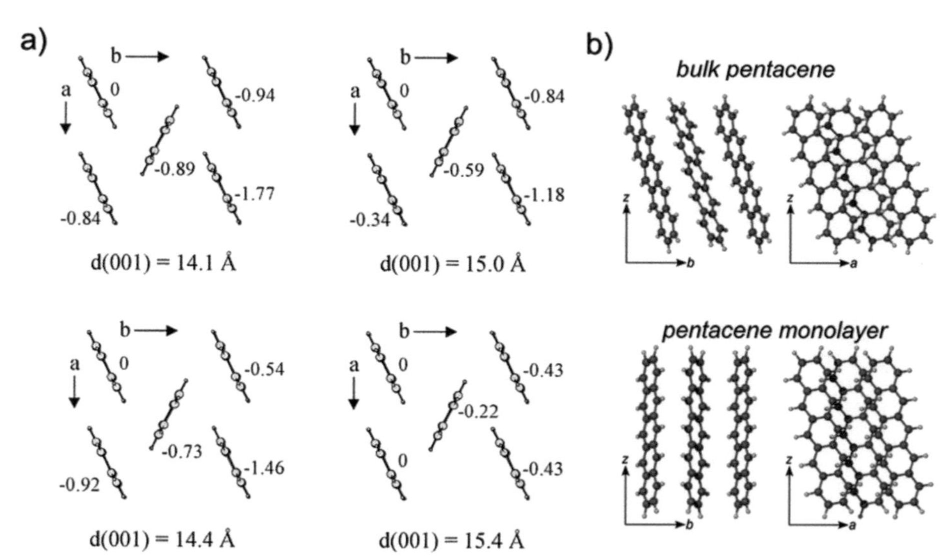

Figure 2.
(a) Schematic drawing of the crystal structures of the pentacene polymorphs [31]. Copyright 2003, American Chemical Society. (b) Normal views of the ab planes of the bulk and the monolayer structures of pentacene [60]. Copyright 2004, American Chemical Society.

reported that the polymorph with d-spacing of 16.2 Å has a mobility of up to 30.6 $cm^2 V^{-1} s^{-1}$ [93]. As for the other polymorphs, to stabilize and isolate a pure polymorph for transistor fabrication has been challenging. Thus, more investigation on the relationship between molecular packing and charge transport for pentacene remains a serious topic of research.

3.2.2 6,13-Bis(triisopropylsilylethynyl)-pentacene

In 2001, Anthony et al. introduced triisopropylsilylethynyl (TIPS) group to the pentacene core, obtaining a very soluble pentacene derivative, i.e. TIPS-pentacene [46]. Different from the herringbone-stacked pentacene molecules, the TIPS-pentacene molecules adopt a brick-wall stacking in solid state. Several recent investigations have revealed that TIPS-pentacene exhibits polymorphism. For instance, Diao et al. fabricated five different polymorphs of TIPS-pentacene by using the solution shearing technology [15]. The five polymorphs have been categorized into three families: I and Ib, II and IIb, and III. Within each family, there is only a slight change in one or two unit cell parameters between the polymorphs. Among different family (polymorphs I, II, and III), the main structural differences are changes in the π-π stacking distance and the extent of overlap between adjacent molecules (**Figure 3**). Form I has larger π-π stacking distance than that of forms II and III, where pair I in form III exhibits a record low-stacking distance of less than 3 Å. As a result, form I exhibits ambipolar charge transport property, with hole and electron mobility of 3.8 and 6.81 $cm^2 V^{-1} s^{-1}$, respectively. Form II shows the highest mobility up to 11 $cm^2 V^{-1} s^{-1}$. In sharp contrast, form III possesses the lowest mobility around 0.09 $cm^2 V^{-1} s^{-1}$. The quantum chemical calculations indicate a much smaller hole transfer integral for form III compared to form I and form II. Form II with moderate π-π stacking distance has the largest hole transfer integral

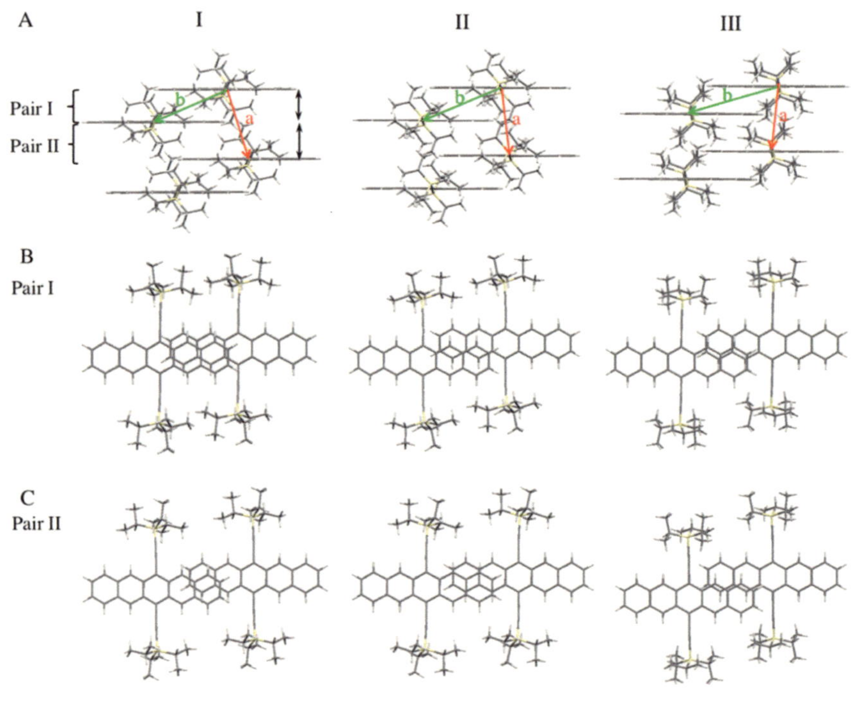

Figure 3.
Comparison of the three major polymorphs of TIPS-pentacene in their π-π stacking (A) and molecular offset along the conjugated backbone (B, C) as obtained from the crystallographic refinement calculations [15]. Copyright 2014, American Chemical Society.

and therefore exhibits the highest mobility. This result shows that small π-π stacking distance does not always help to improve charge transport properties.

3.2.3 2,7-Dioctyl[1]benzothieno[3,2-b][1]benzothiophene

2,7-Dioctyl[1]benzothieno[3,2-b][1]benzothiophene (C8-BTBT) is an extensive studied air-stable organic semiconductor, which often shows very high hole mobility in OFET devices [4, 96, 97]. In a recent study by Yuan et al., ultra-high mobility up to 43 $cm^2 V^{-1} s^{-1}$ was obtained from thin-film transistors based on a metastable polymorph of C8-BTBT film [4]. The metastable polymorph was fabricated by introducing polystyrene additive and using an off-center spin-coating method. After thermal annealing, the metastable polymorph was relaxed to the equilibrium polymorph, along with a sharp decrease of carrier mobility. However, the correlation between molecular packing and charge transport for the C8-BTBT films is difficult to establish. The authors stated that the beam damage during grazing incidence X-ray diffraction (GIXD) measurements made it impossible to obtain the precise crystal packing structure for the metastable polymorph. Moreover, the crystal alignment was also disrupted after thermal annealing, making it difficult to attribute the mobility drop entirely to crystal structure change.

3.2.4 5,11-Bis(triethylsilylethynyl)anthradithiophene

5,11-Bis(triethylsilylethynyl)anthradithiophene (TES-ADT) is a high-performance organic semiconductor with good solubility [53]. In a study by Yu and colleagues, four polymorphs of TES-ADT was obtained in thin films, including three thin-film forms (α, β, and γ polymorphs) and one amorphous form [98]. The α-phase film exhibited the highest hole mobility of 0.4 $cm^2 V^{-1} s^{-1}$, which was about two orders of magnitude higher than that of the β and γ polymorphs. However, in a study by Chen et al., the β-phase film fabricated from toluene by drop casting had a higher mobility (0.22 $cm^2 V^{-1} s^{-1}$) than that of α-phase film from tetrahydrofuran (0.06 $cm^2 V^{-1} s^{-1}$) [52]. The opposite results from these two investigations reveal that to directly correlate mobility to molecular packing from thin-film transistor with different polymorphs is challenging. It is known that many factors can affect the charge transport in thin-film transistors, including film morphology, degree of crystallinity, orientation, grain size, grain boundaries, and so on.

3.3 Single-crystal transistors with different polymorphs

Single-crystal transistors are preferred for fundamental studies on structure-charge transport relationships owning to their high molecular ordering and no grain boundaries. However, structure-property investigations have been successfully performed for only a small number of organic semiconductors by means of single-crystal transistors. Compared to thin-film transistors, the manufacture of single-crystal OFETs is more complicated, which generally requires the use of high-precision deposition or micromanipulation techniques. Moreover, the preparation of single crystals with different polymorphs is very difficult. In this section, some examples of single-crystal OFETs are introduced.

3.3.1 Rubrene

Rubrene is an excellent organic semiconductor with single-crystal mobility up to 20 $cm^2 V^{-1} s^{-1}$ [99]. Four polymorphs of rubrene, including a monoclinic, a triclinic, and two orthorhombic forms, have been known for a long time [30], but the structure-charge transport relationship has only been discussed recently. In an investigation by Matsukawa et al., an orthorhombic single crystal exhibited high carrier mobility up to 1.6 $cm^2 V^{-1} s^{-1}$, while that of the triclinic form was only 0.1 $cm^2 V^{-1} s^{-1}$ [29, 100].

The two polymorphs share similar π-π stacking distances, while the density of the π-stacking column in the (001) plane of orthorhombic crystal structure (**Figure 4a**) is much higher than that of the (0−11) plane of the triclinic crystal structure (**Figure 4b**). In other words, the π-π overlap along the carrier conduction direction in orthorhombic crystal is significantly larger than that in the triclinic crystal. Consequently, the orthorhombic crystal exhibits higher carrier mobility than the triclinic crystal.

3.3.2 Tetrathiafulvalene

Two polymorphs of tetrathiafulvalene (TTF), including a monoclinic orange crystal and a triclinic yellow crystals, were found many years ago [101]. The single-crystal transistors of the two polymorphs were fabricated by Jiang and colleagues recently [102]. Based on single-crystal transistors, the charge transport performance associated with molecular packing was discussed. The monoclinic α-form yielded a higher mobility of approximately 1.2 $cm^2 V^{-1} s^{-1}$, while that of the triclinic β-form was around 0.23 $cm^2 V^{-1} s^{-1}$. Compared to the β-form, the α-form shows strong π-π stacking along the short b axis and short contacts between S atoms (**Figure 5**), which may be the main factor contributing to the higher mobility.

3.3.3 Dibenzotetrathiafulvalene

Brillante et al. investigated the polymorphism of dibenzotetrathiafulvalene (DB-TTF), and four polymorphs were found [49]. The pure α-phase crystals were obtained from chlorobenzene or dimethylformamide solution [103]. In contrast, the β-phase crystals were synthesized from hot saturated toluene solution and were usually accompanied with the α-phase crystals. Moreover, γ-phase thin films were fabricated by ultra-high vacuum (UHV) vapor deposition on SiO_x substrates at temperatures of 50–70°C and drop casting of colloidal composite solutions of polystyrene (PS) and DB-TTF. The fourth polymorph, the δ-phase, was obtained physically pure by crystallization from a mixture solution of isopropanol and nitromethane, as well as by vapor deposition under vacuum. However, among the four polymorphs, only the full-unit-cell parameters of the α- and β-phase crystals have been obtained (**Figure 6**).

In the α-phase crystal, the molecules adopt a face-to-face herringbone structure, possessing a good π-π overlap. Though the β-phase crystal also shows a herringbone

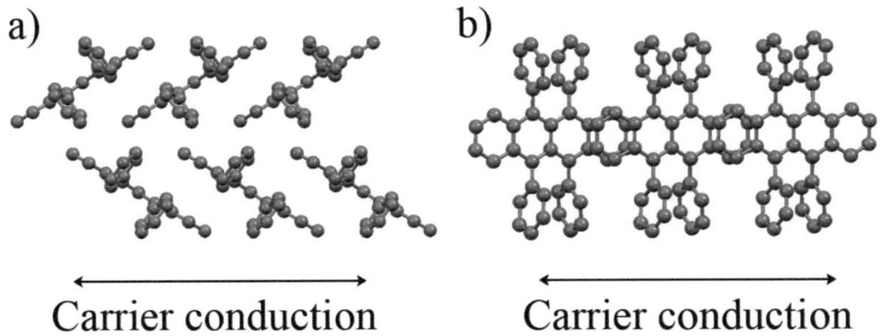

Figure 4.
Crystallographic structures of (a) the orthorhombic crystal (CCDC, 1025043) as viewed as the (001) plane and (b) the triclinic crystal (CCDC, 991020) as viewed as the (0−11) plane.

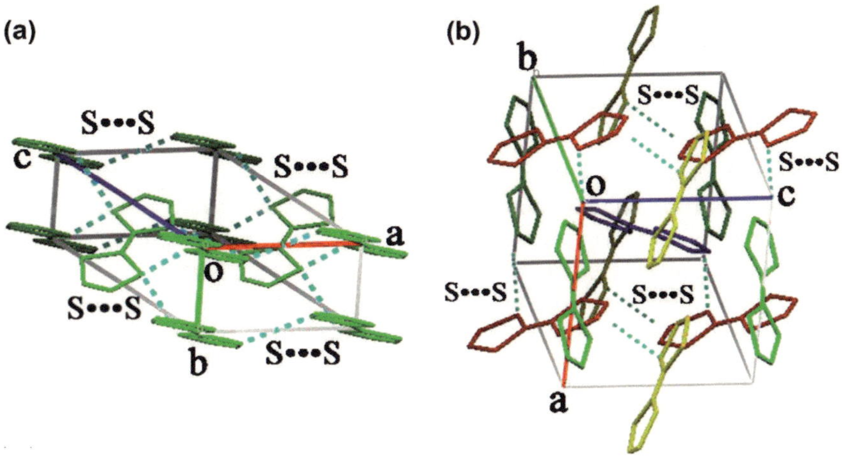

Figure 5.
Crystal packing of (a) α-TTF and (b) β-TTF with S···S interaction [102]. Copyright 2007, American Institute of Physics.

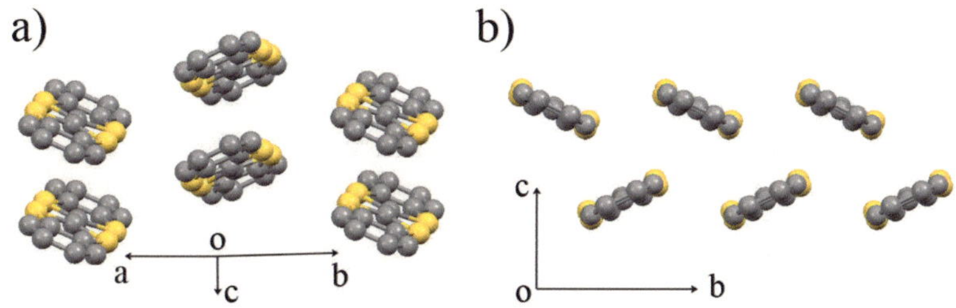

Figure 6.
The molecular packing of (a) the α-phase (CCDC, 1111519) and (b) the β-phase (CCDC, 696271) crystals of DB-TTF.

motif, the edge-to-face molecular packing neither results in π-π overlap between consecutive molecules in the stacks along b axis nor short contacts between adjacent columns. The solution-prepared single-crystal transistors based on α-phase crystal showed best hole mobility of up to 1.0 $cm^2\ V^{-1}\ s^{-1}$ [103].

3.3.4 Dithiophene-tetrathiafulvalene

Dithiophene-tetrathiafulvalene (DT-TTF) is a promising high-performance organic semiconductor, whose single-crystal OFETs were reported with high hole mobility up to 3.6 $cm^2\ V^{-1}\ s^{-1}$ [104]. Long plated crystals of DT-TTF can be easily prepared from a variety of solutions [104–106], which were named as the α-phase. In a study by Pfattner and colleagues, a new β-phase polymorph of DT-TTF was obtained as hexagonal-shaped platelet-like crystals [50]. The β-phase crystals were grown on some substrates from a solution of toluene or dichlorobenzene, mixed with crystals of the α-phase by ultrasonication of the solution before drop casting. The relative ratio of β-DT-TTF increased, whereas the α-phase was mostly obtained in the presence of small seed crystals. This indicates that the crystallization of α-phase probably starts in the solution, while the β-phase crystallizes directly on the substrate. The single-crystal OFETs of the two polymorphs were fabricated, and the device performances were measured, giving the hole mobilities of 1.18 and 0.16 $cm^2\ V^{-1}\ s^{-1}$ from the α-phase and β-phase crystals, respectively. Though the precise crystal structure of the β-phase crystal was not obtained, the analysis of crystal structures for the two polymorphs revealed smaller distances between neighboring molecules in the α-phase crystal which may facilitate charge transport.

3.3.5 7,14-Bis((trimethylsilyl)ethynyl)-dibenzo[b,def]-chrysene

Stevens et al. found two polymorphs of 7,14-bis((trimethylsilyl)ethynyl)-dibenzo[b,def]-chrysene (TMS-DBC) using the physical vapor transport technology [58]. The first polymorph was obtained as red needles at low temperature, which was named as LT-phase (**Figure 7a**). The second polymorph was formed at high temperature as yellow plates and named HT-phase (**Figure 7b**). Further investigations found that the LT-phase can also be fabricated from solution and could not be converted into HT-phase by thermal annealing. Single-crystal OFETs of the two polymorphs were fabricated. The results revealed that the hole mobility of the HT-phase is up to 2.1 $cm^2 V^{-1} s^{-1}$, while that of the LT-phase is only 0.028 $cm^2 V^{-1} s^{-1}$.

As shown in **Figure 7**, the LT-phase adopts one-dimensional (1D) slipped stacking, while the HT-phase exhibits two-dimensional (2D) brick-wall stacking. Quantum chemical calculations revealed that the LT-phase possesses 1D charge transport channel along the π-stacking direction with a transfer integral of -86.8×10^{-3} eV (**Figure 7b**). In contrast, the HT-phase possesses exhibits 2D charge transfer channels with transfer integrals of -77.3×10^{-3} and -41×10^{-3} eV along the t1 and t2 directions, respectively (**Figure 7e**). Though the HT-phase exhibits slightly smaller transfer integral (absolute value) than that of the LT-phase, its 2D charge transport channels benefit charge transfer, which allows charge carriers to take alternative pathways around defects or trap states. As a result, the HT-phase facilitates higher mobility than the LT-phase.

3.3.6 N,N'-bis-(heptafluorobutyl)-2,6-dichloro-1,4,5,8-naphthalene tetracarboxylic diimide

Würthner et al. manufactured single-crystal transistors based on α-phase crystals of N,N′-bis-(heptafluorobutyl)-2,6-dichloro-1,4,5,8-naphthalene tetracarboxylic diimide (Cl2-NDI), which showed electron mobility up to 8.6 $cm^2 V^{-1} s^{-1}$,

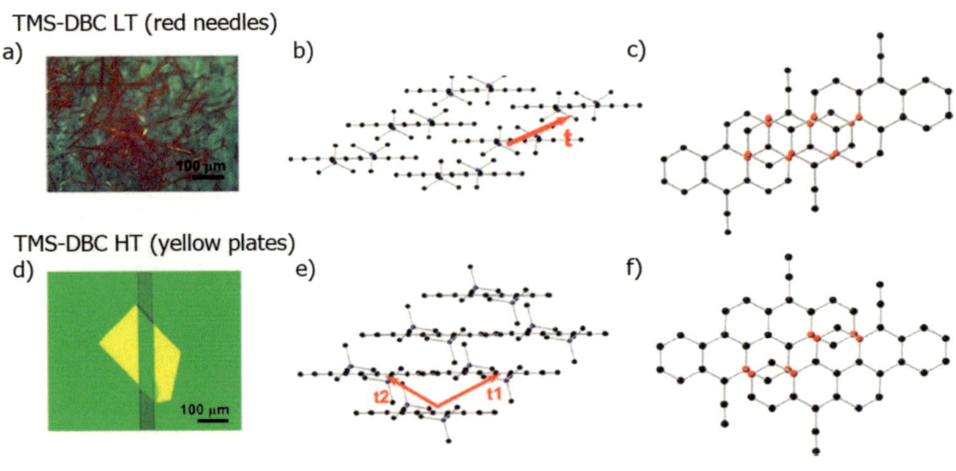

Figure 7.
(a) and (d) Crystal color, size, and shape of the LT red and HT yellow polymorphs of TMS-DBC. (b) and (c) Side and top views of the crystal packing in the red LT polymorph. (e) and (f) Side and top views of the crystal packing in the yellow HT polymorph. The directions corresponding to the largest calculated electronic couplings are indicated with arrows [58]. Copyright 2015, American Chemical Society.

achieving the best performance for air-stable n-type OFETs reported till now [107]. The α-phase polymorph of Cl2-NDI was grown on n-octadecyl triethoxysilane-modified substrates resulting in ribbon-shaped crystals by drop casting via $CHCl_3$ solution. In a study by He et al., the ribbon-shaped β-phase crystals were grown on various substrates like Si/SiO2 by sublimation in air ambient pressure [108]. The single-crystal transistors of β-phase crystals exhibited a maximum electron mobility of 3.5 $cm^2\ V^{-1}\ s^{-1}$.

In the α-phase crystal, Cl2-NDI molecules adopt a herringbone packing motif, where nearly half of each molecular skeleton overlaps with the adjacent molecule at a close π-stack distance of 3.27 Å. In contrast, molecules in the β-phase crystal exhibit a two-dimensional brick-wall packing arrangement with a π-stack distance of 3.29 and 3.32 Å. The selected area electron diffraction (SAED) studies revealed that single-crystal transistors for both the α-phase and β-phase crystals were measured along the π-π stacking direction. Compared to the α-phase crystal, the β-phase crystal possesses a weaker electronic coupling along the π-π stacking direction and a longer percolation pathway for electrons to cross the unit cell, which explains lower carrier mobility.

3.3.7 Dihexyl-dibenzo[d, d']thieno[3, 2-b; 4,5-b']dithiophene

In an investigation by He et al., dihexyl-dibenzo[d, d']thieno[3, 2-b; 4,5-b'] dithiophene (C6-DBTDT) was synthesized efficiently [47]; two polymorphs of C6-DBTDT were obtained by drop casting of solutions with different concentration in chlorobenzene or toluene. The platelet-like α-phase single crystals were prepared through drop casting from a high concentration chlorobenzene solution (5.0 mg/mL). In contrast, the micro-ribbonlike β-phase single crystals were formed from a relatively diluted chlorobenzene solution (0.3 mg/mL). Single-crystal transistors were fabricated, where the α-phase and β-phase crystals exhibited hole mobilities of 8.5 and 18.9 $cm^2\ V^{-1}\ s^{-1}$, respectively.

It is generally believed that only the HOMO level contributes to hole charge carrier transport from one molecule to another adjacent molecule. The electronic couplings of adjacent molecules in the α-phase and β-phase crystals were calculated by the quantum chemical calculations. The results indicated that the electronic couplings of HOMO between adjacent molecules in α-phase crystal are larger than that in β-phase crystal. On the other hand, the electronic couplings of HOMO demonstrate that the α-phase crystals may facilitate charge transport, which is in opposition to the experimental results. The authors noted that the electronic couplings of (HOMO-1)s for β-phase could be much larger than that of α-phase, which is near to the HOMO level. It was thought that (HOMO-1) level plays an important role in the charge transporting behavior.

3.3.8 2,8-Bis(butyl(methyl)amino)-indeno[1,2-b]fluorene-6,12-dione (BMA-IFD)

In a recent investigation by Fan et al., a new molecule, indeno[1,2-b]fluorene-6,12-dione derivative, i.e., BMA-IFD, was designed and synthesized [55]. Two polymorphs of BMA-IFD were easily obtained by crystallization from solution. The ribbon-shaped α-phase crystal (**Figure 8a**) was obtained from chloroform (1 mg/ml), while the flake-shaped β-phase crystal (**Figure 8b**) was obtained from xylene solution (0.2 mg/ml). Single-crystal OFETs were fabricated and measured. The results showed that the β-phase polymorph exhibits hole mobility up to 1.26 $cm^2\ V^{-1}\ s^{-1}$, much higher than that of the α-phase crystals (0.21 $cm^2\ V^{-1}\ s^{-1}$).

The α-phase and the β-phase crystals, whose structures were experimentally determined, were investigated by quantum chemical calculations to associate the

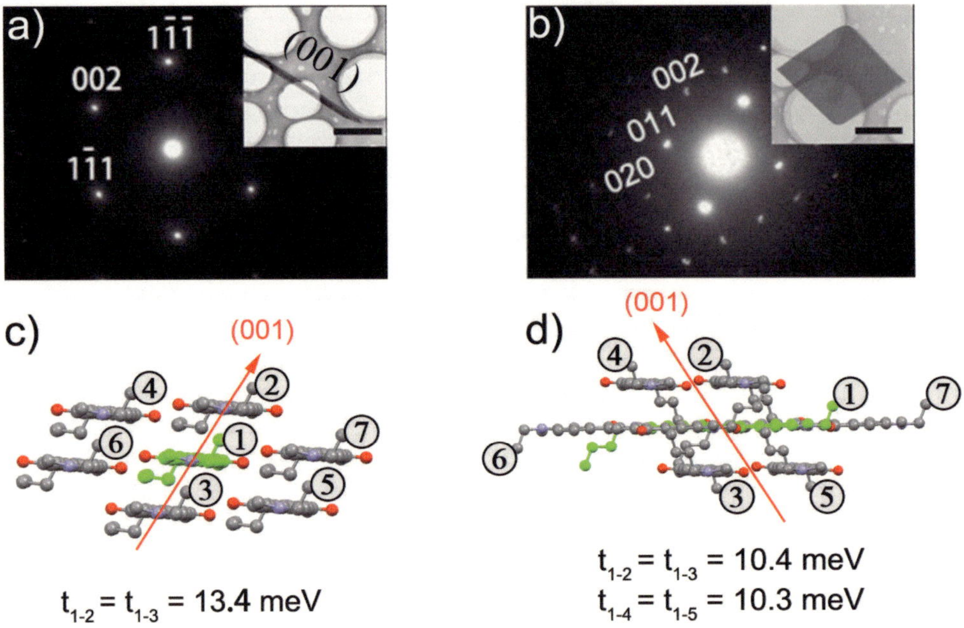

Figure 8.
SAED and TEM images of (a) the α-phase and (b) the β-phase crystals (the scale bar is 5 μm). The transfer integrals of (c) the α-phase and (d) the β-phase crystals along the (001) directions. The molecules in panels e and f are colored differently only for clarity purposes [55]. Copyright 2018, American Chemical Society.

charge transport properties with the molecular packing structures. The results show that a one-dimensional (1D) electron coupling between adjacent molecules is observed in the α-phase crystal (**Figure 8c**). In comparison, a two-dimensional (2D) electron coupling between adjacent molecules is found in the *β*-phase crystal (**Figure 8d**). Though the values of transfer integrals are close for the two polymorphs, the β-phase polymorph possesses a 2D charge transport network and therefore exhibits higher carrier mobility.

3.3.9 Titanyl phthalocyanine

Titanyl phthalocyanine (TiOPC) is a well-known organic semiconductor and photoconductor; however, it exhibits poor solubility in common solvents. In a recent study by Zhang et al., TiOPC crystals were synthesized by physical vapor transport (PVT) technique through a two-zone horizontal tube furnace [6]. Some sheet crystals were obtained at the temperature zone of about 210°C, while some ribbon crystals were grown at the temperature zone of about 180°C. The sheet and ribbon crystals belong to the *α*-phase and *β*-phase polymorphs, respectively. The measurements on single-crystal OFETs of the two polymorphs demonstrated that the *α*-phase crystals exhibit excellent charge transport property with mobility up to 26.8 $cm^2\ V^{-1}\ s^{-1}$, while that of *β*-phase crystals are only 0.1 $cm^2\ V^{-1}\ s^{-1}$.

The crystal structures of the two polymorphs were determined, where the *α*-phase and *β*-phase crystals exhibit a 2D lamellar brick stone motif and an unusual 3D framework, respectively. The main difference in electronic coupling of the two polymorphs was a strong interlayer electronic couplings perpendicular to the current direction in the *β*-phase crystal. The strong interlayer electronic couplings may result in destructive interference effects that remarkably diminish the charge carrier mobility.

4. Conclusion and outlook

Herein, the polymorphism in organic semiconductors is introduced, including the common strategies for polymorph control and investigations on OFETs from different polymorphs. Polymorphism is proved to be an excellent platform to directly correlate the molecular packing with charge transport for organic semiconductors; such investigations are very limited so far. A main challenge is to precisely tailor thermodynamic and kinetic factors of crystal nucleation and growth for large-area thin films or high-quality single crystals. Among the investigations on polymorphism, several polymorphs with outstanding charge transport performance have been obtained, demonstrating that altering the crystal polymorph structure of organic semiconductors is an efficient strategy to access high-performance OFETs. However, the majority of the high-mobility polymorphs are metastable. Consequently, getting insight into the relationship between molecular structure and crystal polymorph remains an important issue, which is essential for the rational design of molecular structures to further develop the desired crystal polymorphs with outstanding electrical characteristics.

Acknowledgements

This work is supported by the National Key R&D Program of China (2017YFA0204903), National Natural Science Foundation of China (NSFC. 51733004, 51525303, 221702085, 21673106, 21602093, 21572086, 1522203), 111 Project, and the Fundamental Research Funds for the Central Universities. The authors thank beam line BL14B1 (Shanghai Synchrotron Radiation Facility) for providing the beam time.

Conflict of interest

The authors declare no conflict of interest.

Author details

Zhi-Ping Fan[1,2] and Hao-Li Zhang[1]*

1 State Key Laboratory of Applied Organic Chemistry, College of Chemistry and Chemical Engineering, Lanzhou University, Lanzhou, China

2 School of Materials Science and Engineering, Nanchang University, Nanchang, China

*Address all correspondence to: haoli.zhang@lzu.edu.cn

References

[1] Dimitrakopoulos CD, Malenfant PRL. Organic thin film transistors for large area electronics. Advanced Materials. 2002;**14**:99-117

[2] Horowitz G. Organic field-effect transistors. Advanced Materials. 1998;**10**:365-377

[3] Lee J et al. Boosting the Ambipolar performance of solution—Processable polymer semiconductors via hybrid side-chain engineering. Journal of the American Chemical Society. 2013;**135**:9540-9547

[4] Yuan Y et al. Ultra-high mobility transparent organic thin film transistors grown by an off-centre spin-coating method. Nature Communications. 2014;**5**:3005

[5] Liu J et al. High mobility emissive organic semiconductor. Nature Communications. 2015;**6**:10032

[6] Zhang Z et al. The impact of interlayer electronic coupling on charge transport in organic semiconductors: A case study on titanylphthalocyanine single crystals. Angewandte Chemie, International Edition. 2016;**55**:5206-5209

[7] Ito Y et al. Crystalline ultrasmooth self-assembled monolayers of alkylsilanes for organic field-effect transistors. Journal of the American Chemical Society. 2009;**131**:9396-9404

[8] DiBenedetto SA et al. Charge conduction and breakdown mechanisms in self-assembled nanodielectrics. Journal of the American Chemical Society. 2009;**131**:7158-7168

[9] Asadi K et al. Single-layer pentacene field-effect transistors using electrodes modified with self-assembled monolayers. Advanced Materials. 2009;**21**:4109-4114

[10] Jones AOF et al. Substrate-induced phase of a 1 benzothieno 3,2-b benzothiophene derivative and phase evolution by aging and solvent vapor annealing. ACS Applied Materials & Interfaces. 2015;**7**:1868-1873

[11] Di C-A et al. Solvent-assisted re-annealing of polymer films for solution-processable organic field-effect transistors. Advanced Materials. 2010;**22**:1273-1277

[12] Kim Y et al. π-Conjugated polymers incorporating a novel planar quinoid building block with extended delocalization and high charge carrier mobility. Advanced Materials. 2018;**30**:1706557

[13] Matsidik R et al. Defect-free naphthalene diimide bithiophene copolymers with controlled molar mass and high performance via direct arylation polycondensation. Journal of the American Chemical Society. 2015;**137**:6705-6711

[14] Giri G et al. Tuning charge transport in solution-sheared organic semiconductors using lattice strain. Nature. 2011;**480**:504-508

[15] Diao Y et al. Understanding polymorphism in organic semiconductor thin films through nanoconfinement. Journal of the American Chemical Society. 2014;**136**: 17046-17057

[16] Purdum GE et al. Understanding polymorph transformations in core-chlorinated naphthalene diimides and their impact on thin-film transistor performance. Advanced Functional Materials. 2016;**26**:2357-2364

[17] Chung H, Diao Y. Polymorphism as an emerging design strategy for high performance organic electronics. Journal of Materials Chemistry C. 2016;**4**:3915-3933

[18] He T et al. Single-crystal field-effect transistors of new Cl2-NDI polymorph processed by sublimation in air. Nature Communications. 2015;**6**:5954

[19] Kuhnert-Brandstätter M, Sollinger HW. Thermal analytical and infrared spectroscopic investigations on polymorphic organic compounds—V. Microchimica Acta. 1989;**99**:125-136

[20] Threlfall TL. Analysis of organic polymorphs: A review. The Analyst. 1995;**120**:2435-2460

[21] Burger A, Ramberger R. On the polymorphism of pharmaceuticals and other molecular crystals. II. Microchimica Acta. 1979;**72**:273-316

[22] Llinàs A, Goodman JM. Polymorph control: Past, present and future. Drug Discovery Today. 2008;**13**:198-210

[23] Dou JH et al. Fine-tuning of crystal packing and charge transport properties of BDOPV derivatives through fluorine substitution. Journal of the American Chemical Society. 2015;**137**:15947-15956

[24] Fan Z-P et al. Boosting the charge transport property of indeno[1,2-b] fluorene-6,12-dione though incorporation of sulfur- or nitrogen-linked side chains. Advanced Functional Materials. 2017;**27**:1702318

[25] Lei T, Wang J-Y, Pei J. Roles of flexible chains in organic semiconducting materials. Chemistry of Materials. 2014;**26**:594-603

[26] Mei J et al. Siloxane-terminated solubilizing side chains: Bringing conjugated polymer backbones closer and boosting hole mobilities in thin-film transistors. Journal of the American Chemical Society. 2011;**133**:20130-20133

[27] Huang F et al. Development of new conjugated polymers with donor–π-bridge–acceptor side chains for high performance solar cells. Journal of the American Chemical Society. 2009;**131**:13886-13887

[28] Bergantin S, Moret M. Rubrene polymorphs and derivatives: The effect of chemical modification on the crystal structure. Crystal Growth & Design. 2012;**12**:6035-6041

[29] Matsukawa T et al. Growth of thin rubrene single crystals from 1-propanol solvent. Journal of Crystal Growth. 2010;**312**:310-313

[30] Jurchescu OD, Meetsma A, Palstra TTM. Low-temperature structure of rubrene single crystals grown by vapor transport. Acta Crystallographica Section B. 2006;**62**: 330-334

[31] Mattheus CC et al. Modeling the polymorphism of pentacene. Journal of the American Chemical Society. 2003;**125**:6323-6330

[32] Schiefer S et al. Determination of the crystal structure of substrate-induced pentacene polymorphs in fiber structured thin films. Journal of the American Chemical Society. 2007;**129**:10316-10317

[33] Pithan L et al. Light controls polymorphism in thin films of sexithiophene. Crystal Growth & Design. 2015;**15**:1319-1324

[34] Moser A et al. A disordered layered phase in thin films of sexithiophene. Chemical Physics Letters. 2013;**574**:51-55

[35] Brillante A et al. Confocal Raman spectroscopy of α-sexithiophene: From bulk crystals to field-effect transistors. Advanced Functional Materials. 2007;**17**:3119-3127

[36] Servet B et al. Polymorphism and charge transport in vacuum-evaporated sexithiophene films. Chemistry of Materials. 1994;**6**:1809-1815

[37] Takimiya K et al. Organic semiconductors based on [1] benzothieno[3,2-b][1]benzothiophene substructure. Accounts of Chemical Research. 2014;**47**:1493-1502

[38] Gbabode G et al. X-ray structural investigation of nonsymmetrically and symmetrically alkylated 1 benzothieno 3,2-b benzothiophene derivatives in bulk and thin films. ACS Applied Materials & Interfaces. 2014;**6**:13413-13421

[39] Xue G et al. Ambipolar charge transport of TIPS-pentacene single-crystals grown from non-polar solvents. Materials Horizons. 2015;**2**:344-349

[40] Mei J et al. Integrated materials design of organic semiconductors for field-effect transistors. Journal of the American Chemical Society. 2013;**135**:6724-6746

[41] Li H et al. High-mobility field-effect transistors from large-area solution-grown aligned C60 single crystals. Journal of the American Chemical Society. 2012;**134**:2760-2765

[42] Watanabe M et al. The synthesis, crystal structure and charge-transport properties of hexacene. Nature Chemistry. 2012;**4**:574-578

[43] Takeyama Y, Ono S, Matsumoto Y. Organic single crystal transistor characteristics of single-crystal phase pentacene grown by ionic liquid-assisted vacuum deposition. Applied Physics Letters. 2012;**101**:083303

[44] Liu C et al. Solution-processable organic single crystals with bandlike transport in field-effect transistors. Advanced Materials. 2011;**23**:523-526

[45] Zhang X, Dong H, Hu W. Organic semiconductor single crystals for electronics and photonics. Advanced Materials. 2018;**30**:e1801048

[46] Anthony JE et al. Functionalized pentacene: Improved electronic properties from control of solid-state order. Journal of the American Chemical Society. 2001;**123**:9482-9483

[47] He P et al. Tuning the crystal polymorphs of alkyl thienoacene via solution self-assembly toward air-stable and high-performance organic field-effect transistors. Advanced Materials. 2015;**27**:825-830

[48] Hu X et al. In situ polymerization of ethylenedioxythiophene from sulfonated carbon nanotube templates: Toward high efficiency ITO-free solar cells. Journal of Materials Chemistry A. 2016;**4**:6645-6652

[49] Brillante A et al. The four polymorphic modifications of the semiconductor dibenzo-tetrathiafulvalene. CrystEngComm. 2008;**10**:1899-1909

[50] Pfattner R et al. High-performance single crystal organic field-effect transistors based on two dithiophene-tetrathiafulvalene (DT-TTF) polymorphs. Advanced Materials. 2010;**22**:4198-4203

[51] Giri G et al. One-dimensional self-confinement promotes polymorph selection in large-area organic semiconductor thin films. Nature Communications. 2014;**5**:3573

[52] Chen J et al. Solvent-type-dependent polymorphism and charge transport in a long fused-ring organic semiconductor. Nanoscale. 2014;**6**:449-456

[53] Payne MM et al. Organic field-effect transistors from solution-deposited functionalized acenes with mobilities as high as 1 cm^2/V s. Journal of the American Chemical Society. 2005;**127**:4986-4987

[54] Lee SS et al. Controlling nucleation and crystallization in solution-processed

organic semiconductors for thin-film transistors. Advanced Materials. 2009;**21**:3605-3609

[55] Fan ZP et al. Enhancing the thermal stability of organic field-effect transistors by electrostatically interlocked 2D molecular packing. Chemistry of Materials. 2018;**30**: 3638-3642

[56] Chen J, Trout BL. A computational study of the mechanism of the selective crystallization of α- and β-glycine from water and methanol–water mixture. The Journal of Physical Chemistry B. 2010;**114**:13764-13772

[57] Laudise RA et al. Physical vapor growth of organic semiconductors. Journal of Crystal Growth. 1998;**187**: 449-454

[58] Stevens LA et al. Temperature-mediated polymorphism in molecular crystals: The impact on crystal packing and charge transport. Chemistry of Materials. 2015;**27**:112-118

[59] Wedl B et al. Crystallisation kinetics in thin films of dihexyl-terthiophene: The appearance of polymorphic phases. RSC Advances. 2012;**2**:4404-4414

[60] Fritz SE et al. Structural characterization of a pentacene monolayer on an amorphous SiO_2 substrate with grazing incidence X-ray diffraction. Journal of the American Chemical Society. 2004;**126**:4084-4085

[61] Mannsfeld SCB et al. Precise structure of pentacene monolayers on amorphous silicon oxide and relation to charge transport. Advanced Materials. 2009;**21**:2294-2298

[62] Campbell RB, Robertson JM, Trotter J. The crystal structure of hexacene, and a revision of the crystallographic data for tetracene. Acta Crystallographica. 1962;**15**:289-290

[63] Mattheus CC et al. Identification of polymorphs of pentacene. Synthetic Metals. 2003;**138**:475-481

[64] Siegrist T et al. Enhanced physical properties in a pentacene polymorph. Angewandte Chemie, International Edition. 2001;**40**:1732-1736

[65] Jones AOF et al. Substrate-induced and thin-film phases: Polymorphism of organic materials on surfaces. Advanced Functional Materials. 2016;**26**:2233-2255

[66] Chen J et al. Conjugated polymer-mediated polymorphism of a high performance, small-molecule organic semiconductor with tuned intermolecular interactions, enhanced long-range order, and charge transport. Chemistry of Materials. 2013;**25**: 4378-4386

[67] Hiszpanski AM, Loo Y-L. Directing the film structure of organic semiconductors via post-deposition processing for transistor and solar cell applications. Energy & Environmental Science. 2014;**7**:592-608

[68] Ullah Khan H et al. Solvent vapor annealing in the molecular regime drastically improves carrier transport in small-molecule thin-film transistors. ACS Applied Materials & Interfaces. 2013;**5**:2325-2330

[69] Khim D et al. High performance and stable N-channel organic field-effect transistors by patterned solvent-vapor annealing. ACS Applied Materials & Interfaces. 2013;**5**:10745-10752

[70] Hiszpanski AM et al. Tuning polymorphism and orientation in organic semiconductor thin films via post-deposition processing. Journal of the American Chemical Society. 2014;**136**:15749-15756

[71] Campione M et al. Crystal-to-crystal phase transition in alpha-quaterthiophene: An optical and

structural study. Journal of Applied Physics. 2007;**101**:083512

[72] Bergantin S et al. Pressure-induced conformational change in organic semiconductors: Triggering a reversible phase transition in rubrene. Journal of Physical Chemistry C. 2014;**118**:13476-13483

[73] Iwasa Y et al. New phases of C60 synthesized at high pressure. Science. 1994;**264**:1570-1572

[74] Yoo CS, Nellis WJ. Phase-transformations in carbon fullerenes at high shock pressures. Science. 1991;**254**: 1489-1491

[75] Wang L et al. Synthesis and high pressure induced amorphization of C60 nanosheets. Applied Physics Letters. 2007;**91**:103112

[76] Wang L et al. Long-range ordered carbon clusters: A crystalline material with amorphous building blocks. Science. 2012;**337**:825-828

[77] Ito H et al. Mechanical stimulation and solid seeding trigger single-crystal-to-single-crystal molecular domino transformations. Nature Communications. 2013;**4**:2009

[78] Sagara Y, Kato T. Mechanically induced luminescence changes in molecular assemblies. Nature Chemistry. 2009;**1**:605-610

[79] Lee S-L et al. Flow-assisted 2D polymorph selection: Stabilizing metastable monolayers at the liquid–solid interface. Journal of the American Chemical Society. 2014;**136**: 7595-7598

[80] Bernstein J. Crystal-growth, polymorphism and structure-property relationships in organic-crystals. Journal of Physics D: Applied Physics. 1993;**26**:B66-B76

[81] Holstein T. Studies of polaron motion: Part I. The molecular-crystal model. Annals of Physics. 1959;**8**: 325-342

[82] Holstein T. Studies of polaron motion: Part II. The “small” polaron. Annals of Physics. 1959;**8**:343-389

[83] Coropceanu V et al. Charge transport in organic semiconductors. Chemical Reviews. 2007;**107**:926-952

[84] Mas-Torrent M, Rovira C. Role of molecular order and solid-state structure in organic field-effect transistors. Chemical Reviews. 2011;**111**: 4833-4856

[85] Sakanoue T, Sirringhaus H. Band-like temperature dependence of mobility in a solution-processed organic semiconductor. Nature Materials. 2010;**9**:736-740

[86] Marcus RA. Electron transfer reactions in chemistry: Theory and experiment. Reviews of Modern Physics. 1993;**65**:599-610

[87] Brédas J-L et al. Charge-transfer and energy-transfer processes in π-conjugated oligomers and polymers: A molecular picture. Chemical Reviews. 2004;**104**:4971-5004

[88] Bredas JL et al. Organic semiconductors: A theoretical characterization of the basic parameters governing charge transport. PNAS. 2002;**99**:5804-5809

[89] Tobjörk D, Österbacka R. Paper electronics. Advanced Materials. 2011;**23**:1935-1961

[90] Jeon PJ et al. Ultrasensitive low power-consuming strain sensor based on complementary inverter composed of organic p- and n-channels. Organic Electronics. 2016;**32**:208-212

[91] Zhang F et al. Flexible and self-powered temperature-pressure

dual-parameter sensors using microstructure-frame-supported organic thermoelectric materials. Nature Communications. 2015;**6**:8356

[92] Mills WH, Mills M. CCXXX.—The synthetical production of derivatives of dinaphthanthracene. Journal of the Chemical Society, Transactions. 1912;**101**:2194-2208

[93] Ji D et al. Surface polarity and self-structured nanogrooves collaboratively oriented molecular packing for high crystallinity toward efficient charge transport. Journal of the American Chemical Society. 2017;**139**:2734-2740

[94] Kakudate T, Yoshimoto N, Saito Y. Polymorphism in pentacene thin films on SiO_2 substrate. Applied Physics Letters. 2007;**90**:081903

[95] Jurchescu OD et al. Interface-controlled, high-mobility organic transistors. Advanced Materials. 2007; **19**:688-692

[96] Minemawari H et al. Inkjet printing of single-crystal films. Nature. 2011;**475**:364-367

[97] Uemura T et al. Very high mobility in solution-processed organic thin-film transistors of highly ordered [1] benzothieno[3,2-b]benzothiophene derivatives. Applied Physics Express. 2009;**2**:111501

[98] Yu L et al. Influence of solid-state microstructure on the electronic performance of 5,11-Bis (triethylsilylethynyl) Anthra-dithiophene. Chemistry of Materials. 2013;**25**:1823-1828

[99] Takeya J et al. Very high-mobility organic single-crystal transistors with in-crystal conduction channels. Applied Physics Letters. 2007;**90**:102120

[100] Matsukawa T et al. Polymorphs of rubrene crystal grown from solution. Japanese Journal of Applied Physics. 2010;**49**:085502

[101] Ellern A et al. A new polymorphic modification of Tetrathiafulvalene. Crystal structure, lattice energy and intermolecular interactions. Chemistry of Materials. 1994;**6**:1378-1385

[102] Jiang H et al. Phase dependence of single crystalline transistors of tetrathiafulvalene. Applied Physics Letters. 2007;**91**:123505

[103] Mas-Torrent M et al. Single-crystal organic field-effect transistors based on dibenzo-tetrathiafulvalene. Applied Physics Letters. 2004;**86**:012110

[104] Leufgen M et al. High-mobility tetrathiafulvalene organic field-effect transistors from solution processing. Organic Electronics. 2008;**9**:1101-1106

[105] Mas-Torrent M et al. High mobility of dithiophene-tetrathiafulvalene single-crystal organic field effect transistors. Journal of the American Chemical Society. 2004;**126**:984-985

[106] Mas-Torrent M et al. Temperature dependence of the electrical properties of single-crystals of dithiophene-tetrathiafulvalene (DT-TTF). Synthetic Metals. 2004;**146**:265-268

[107] He T, Stolte M, Würthner F. Air-stable n-channel organic single crystal field-effect transistors based on microribbons of core-chlorinated naphthalene diimide. Advanced Materials. 2013;**25**:6951-6955

[108] He T et al. Single-crystal field-effect transistors of new Cl(2)-NDI polymorph processed by sublimation in air. Nature Communications. 2015;**6**:5954

Chapter 5

High Capacitance Dielectrics for Low Voltage Operated OFETs

Navid Mohammadian and Leszek A. Majewski

Abstract

Low-voltage, organic field-effect transistors (OFETs) have a high potential to be key components of low-cost, flexible, and large-area electronics. However, to be able to employ OFETs in the next generation of the electronic devices, the reduction of their operational voltage is urgently needed. Ideally, to be power efficient, OFETs are operated with gate voltages as low as possible. To fulfill this requirement, low values of transistor threshold voltage (V_t) and subthreshold swing (*SS*) are essential. Ideally, V_t should be around 0 V and *SS* close to 60 mV/dec, which is the theoretical limit of subthreshold swing at 300 K. This is a very challenging task as it requires the gate dielectric thickness to be reduced below 10 nm. Here, the most promising strategies toward high capacitance dielectrics for low voltage operated OFETs are covered and discussed.

Keywords: thin-film transistor (TFT), organic field-effect transistor (OFET), low voltage transistor operation, high gate capacitance, ultra-thin dielectric, high-k dielectric, high-k/low-k hybrid dielectric, self-assembled monolayer (SAM), anodization

1. Introduction

In this chapter, the most important approaches toward reducing the operating voltage of organic field-effect transistors (OFETs) are described. First, the operation principle of OFETs is covered. This includes the description of the most common organic FET structures and the discussion of the device physics, which has mostly been derived from the theory of the metal-oxide-semiconductor field-effect transistor (MOSFET). Next, the key parameters of organic field-effect transistors that determine the operational voltage of these devices are discussed. Then, compatible electronic materials and device fabrication methods for low voltage operated OFETs are introduced. Finally, the chapter ends by presenting the state-of-the-art low voltage organic transistors and describing the latest key developments relating to manufacturing of such devices.

This chapter is organized in the following order: first, a brief overview of thin-film transistor (TFT) history, applications, device architectures, as well as the basics of TFT operation and differences among TFTs, OFETs, and MOSFETs are presented in Section 2; then, different methods of decreasing the operating voltage of TFTs and OFETs are considered, and various electronic materials and fabrication techniques, which are used to realize low voltage transistor operation, are discussed in Section 3. Lastly, the key findings of this chapter are summarized in conclusions.

2. Thin-film transistors (TFTs)

Thin-film transistors (TFTs) were first introduced by Weimer in 1962 and are a class of field-effect transistors (FETs), which rely on the application of an electric field to the gate electrode to modulate the density of charge carriers in the channel, which is formed at the dielectric/semiconductor interface. A typical FET is made by stacking thin films of a semiconductor layer, a dielectric layer, and metal contacts. Therefore, it behaves in a similar way to the metal-oxide-semiconductor field-effect transistor (MOSFET). Even though TFTs and MOSFETs have been developed simultaneously, MOSFETs have dominated the majority of microelectronic research interests and industrial production due to their much better performance. However, the high MOSFET performance usually comes with high cost and high temperature processing, which is not compatible with the growing demand for low cost, low temperature processing of flexible, stretchable, and large-area electronics. Other dissimilarities between TFTs and MOSFETs, as shown in **Figure 1**, include structural and material differences. Usually, TFTs are made on insulating substrates [e.g. glass or flexible films such as polyethylene terephthalate (PEN) or poly(ethylene terephthalate) (PET)], while MOSFETs are fabricated on semiconducting silicon wafers that simultaneously act as substrates and device active layers. Importantly, both types of transistors operate in a fundamentally different way: MOSFETs operate in the inversion mode, and TFTs operate in the accumulation (enhancement) mode [1, 2].

The demand for low cost, large-area applications in flat panel displays (FPDs) fuelled research on finding a viable substitution for crystalline and polycrystalline silicon. In 1979, significantly cheaper amorphous hydrogenated silicon (a-Si:H) was developed, and subsequently, it was introduced to thin-film transistors as the active layer that resulted in an increased interest in TFTs [3]. Since then, the use of a-Si:H TFTs has gradually grown, and eventually, they have started to dominate the whole liquid crystal display (LCD) industry. Nowadays, a-Si:H TFTs are the backbone of both active matrix liquid crystal displays (AMLCDs) and active matrix organic light-emitting diode (AMOLED) display technologies.

In 1987, Koezuka et al. reported a special type of TFTs that used organic semiconductors as the active layer, so-called organic field-effect transistors (OFETs). The demonstrated devices used electrochemically polymerized polythiophene, which belongs to the family of conducting (i.e. conjugated) polymers (CPs) as the active layer [4]. Accordingly, it has been shown that the thin-film transistor design is the structure of choice for low conductivity materials such as organic semiconductors. As a result, the TFT design was utilized to realize a wide range of field-effect transistors using organic semiconductors (OSCs) [5]. Since then, the performance of organic semiconductors has continuously improved, and

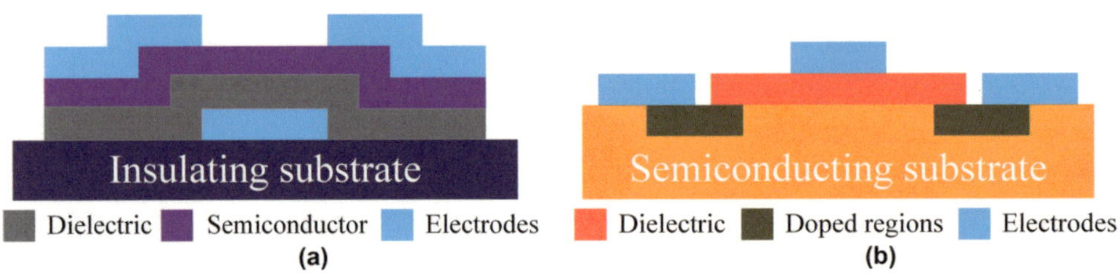

Figure 1.
Typical structures of (a) a bottom-gate top-contact TFT and (b) a MOSFET. Doped regions refer to the MOSFET source and drain regions.

nowadays, OFETs can compete with or even surpass a-Si:H TFTs in a broad range of electronic applications [6].

Although Si is the most common material used in electronics due to its advantages such as abundancy, low cost manufacturing, ease of doping, and high charge carrier mobility, novel applications in electronics such as flexible, stretchable, and large-area circuits and displays have directed the attention of scientists to alternative materials. Frequently, electronic devices such as discrete sensors, simple displays, as well as basic RFID tags and smart cards are realized in various shapes and sizes, and thus, unconventional electronic materials and device processing techniques have had to be researched to appropriately respond to these upward demands. From the most recent trends in material research, it appears that organic and metal oxide semiconductors are the two most promising alternatives to silicon for low cost electronic applications.

In addition, it is believed that many next generation electronic devices will be portable and thus will require considerably reduced power consumption. As a result, the rapid development of low voltage TFTs and OFETs is highly desirable. However, before such low power devices can be realized, a significant reduction in transistor operational voltage is required. Unfortunately, it is still extremely challenging for both organic and metal oxide semiconductor transistors to achieve high performance and low operating voltages at the same time [7–9]. In this chapter, the most promising approaches toward high capacitance dielectrics for high performance, low voltage TFTs, and OFETs are discussed and evaluated.

2.1 OFET architectures

Organic field-effect transistors can be fabricated on a variety of rigid and flexible substrates, namely, silicon/silicon oxide wafers, glass, PEN, PET, and other types of flexible films. Depending on the position of the gate electrode and where each layer is deposited, OFETs are categorized into four different structures, namely, coplanar bottom-gate, staggered bottom-gate, coplanar top-gate, and staggered top-gate, as shown in **Figure 2**. Staggered (also called bottom-contact) or coplanar (also called top-contact) configurations refer to whether the drain/source and gate electrodes are on the opposite or on the same side regarding the semiconductor layer. Although all of the abovementioned structures are used in the fabrication of OFETs, each of them shows a better performance in a particular application, and/or due to the fabrication limitations, one is more desirable than the other [3, 10]. For example, the coplanar top-gate structure is routinely used in flat

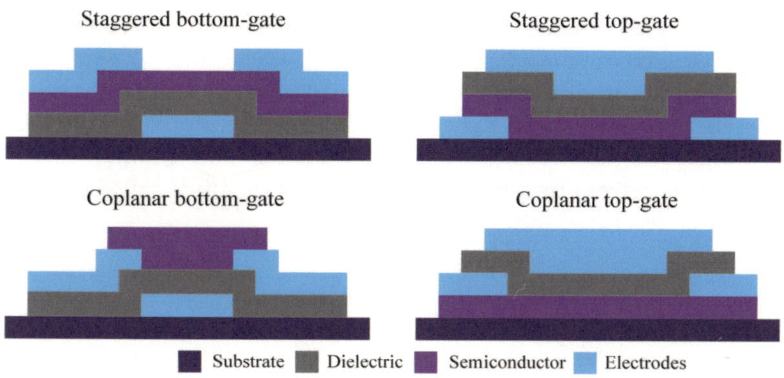

Figure 2.
The most common structures of OFETs.

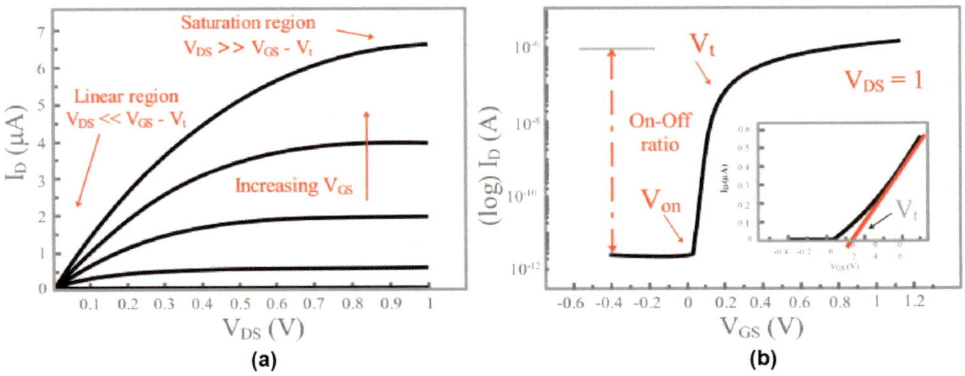

Figure 3.
Typical output (a) and transfer (b) characteristics of a low voltage OFET operating at 1 V.

panel displays. In this structure, the drain/source electrodes are self-aligned with the channel region, which result in the minimization of the parasitic capacitance that reduces image flicker and sticking [10]. Also, top-gate structures usually have higher mobility than their bottom-gate counterparts [11]. The coplanar bottom-gate structure is usually used in electronic nose applications (i.e. e-sensing) due to a larger sensing area [12]. As the device structure directly affects the device performance in terms of contact resistance, parasitic capacitance, charge carrier mobility, sensing capability, and so on, it is crucial to meticulously determine the device architecture regarding the intended use of the device.

2.2 Fundamentals of OFET operation

In an n-channel OFET operating in the accumulation (enhancement) mode, when a positive voltage bias is applied to the gate terminal, electrons start to accumulate at the dielectric/semiconductor interface, which forms a current path (channel) between the source and the drain contacts. Once the source-drain voltage bias is applied, the current starts to flow from the source to the drain electrode. As illustrated in **Figure 3**, depending on the condition of the channel, there are three different operation modes of OFETs, namely, cut-off region, as well as linear and saturation regimes. If the applied gate voltage (V_{GS}) is below a certain value, i.e., smaller than the threshold voltage (V_t), then it is not possible to accumulate enough charges (i.e. electrons) to open the channel. Therefore, no current can flow in the channel, which is called a cut-off region. On the other hand, if $V_{GS} > V_t$, one can have two scenarios: first, when the drain-source voltage (V_{DS}) is equal or larger than V_{GS}-V_t, the amount of the source-drain current (I_{DS}) flowing in the channel is constant and the OFET works in the saturation regime; second, if V_{DS} is lower than V_{GS}-V_t, I_{DS} follows the Ohm's law and the resistance of the channel (R_C) is proportional to V_{DS} and inversely proportional to I_{DS}. This region is called the linear regime, and I_{DS} is described by:

$$I_{DS} = C_G \mu_{lin} \frac{W}{L} [(V_{GS} - V_t) V_{DS}], \tag{1}$$

where C_G is the gate capacitance per unit area, μ_{lin} is the charge carrier mobility in the linear regime, W is the channel width, and L is the channel length of the device.

In the saturation regime, V_{DS} is equal or larger than V_{GS}-V_t, and I_{DS} is independent of V_{DS}. Therefore, the equation is simplified to:

$$I_{DS} = \frac{W}{2L} C_G \mu_{sat} (V_{GS} - V_t)^2, \tag{2}$$

where μ_{sat} is the charge carrier mobility in the saturation regime.

Typical output and transfer characteristics of an OFET operating at 1 V are illustrated in **Figure 3a, b**, respectively. In general, the most essential OFET parameters are as follows (cf. **Figure 3a, b**):

- Turn-on voltage (V_{ON}) is the value of V_{GS} at which I_{DS} starts to increase.
- Threshold voltage (V_t) is the minimum V_{GS} at which the number of the accumulated charge carriers at the dielectric/semiconductor interface is sufficient to create a conduction path (channel) between the source and the drain electrodes. The lower V_t, the lower the operational voltage of an OFET.
- Subthreshold swing (*SS*) is the parameter, which describes the necessary V_{GS} to increase I_{DS} by one order of magnitude (decade) in the subthreshold region, i.e., $V_{ON} < V_{GS} < V_t$. As small value of *SS* as possible is highly desirable because it leads to lower device power consumption and higher device switching speed. It is usually determined by the following expression:

$$SS = \ln 10 \frac{k_b T}{q} \left(1 + \frac{C_{ch}}{C_i}\right), \tag{3}$$

where k_b is the Boltzmann constant, T is the temperature in Kelvin, q is the electron charge, C_i is the gate dielectric capacitance, and C_{ch} is the effective channel capacitance.

- On-off current ratio defines a ratio of the measured maximum to minimum source-drain current. High "on" and low "off" currents in an OFET are highly desirable.
- The field-effect mobility of charge carriers in the linear (μ_{lin}) and saturation regimes (μ_{sat}) is usually determined by calculating the transconductance (i.e., $g_m = \frac{dI_{DS}}{dV_{GS}}$) in both regions, respectively. In the linear regime,

$$\mu_{lin} = \frac{g_m}{C_i \frac{W}{L} V_{DS}}, \tag{4}$$

and in the saturation regime,

$$\mu_{sat} = \frac{\left(\frac{d\sqrt{I_{DS}}}{dV_{GS}}\right)}{\frac{1}{2} C_i \frac{W}{L}} \tag{5}$$

- The total interfacial trap density N_{it} of an OFET can be calculated by Eq. (6) [9]:

$$N_{it} = \left(\frac{SS \log(e)}{kT/q} - 1\right) \frac{C_i}{q^2} \tag{6}$$

where C_i is the gate capacitance density, q is the electron charge, k is the Boltzmann constant, T is the temperature in Kelvin, and *SS* is the subthreshold swing.

3. Low voltage OFETs

As mentioned in the previous section, I_{DS} is described in the linear and saturation regimes by Eqs. (2) and (3), respectively. In the ideal case, the source-drain current should be maximum, while the gate bias voltage is as low as possible. However, this only can be achieved when both the threshold voltage (V_t) and the subthreshold swing (*SS*) are sufficiently low enabling an OFET to be operated at a low voltage and maximum performance [13]. Referring back to Eq. (2), the only parameters that can be changed to compensate the reduction in I_{DS} are the gate dielectric capacitance (C_G) and the channel width (*W*) and length (*L*). However, *W* and *L* depend on the device geometry. Therefore, in order to accumulate the same number of charges within the channel of an OFET and maintain high I_{DS}, it is essential to increase the gate dielectric capacitance (C_G). One may argue that increasing the gate capacitance may deteriorate the transistor's switching speed. Indeed, the maximum switching speed of an OFET is usually defined by its cut-off frequency f_c as shown in Eq. (7) [14]:

$$f_c = \frac{g_m}{2\pi C_G} = \frac{\mu(V_G - V_t)}{2\pi L^2}, \tag{7}$$

where μ_{sat} is the charge carrier mobility in the saturation regime, L is the channel length, V_G is the gate voltage, V_t is the threshold voltage, and f_c is quantified by the g_m/C_G ratio.

As can be seen, increasing the gate capacitance directly decreases the cut-off frequency. However, low operation voltage OFETs are designed for completely different purposes and are typically not used in high switching speed applications because they are not meant to be a substitution for silicon-based transistors. For example, in OFET-based sensors, which generally sense analog quantities such as analyte concentration or pressure, the high operating frequency is not needed as analog quantities do not change rapidly [15]. Therefore, delivering the highest possible source-drain current at the lowest possible gate voltage is more critical than the high operation frequency. However, like in the design of any other electronic systems, some applications require high switching speed. It has been shown that thinning of the channel layer thickness [16] and engineering of the semiconductor/dielectric interface in transistors may result in reduced trap density (N_{it}) within the channel [17], and in consequence, OFETs with improved f_c. However, depending on the intended applications a trade-off situation should always be considered. To better understand the operation physics of organic FETs, the next part of this chapter discusses the background of dielectrics and the theory of parallel plate capacitors, as well as the influence of the gate insulator properties on the performance of OFETs.

3.1 Dielectrics: background

By definition, an insulator is a material, which has an extremely high resistivity to electric current. In other words, a lack of charge transport in insulator materials leads to insulating behavior. In general, insulators can be polar or nonpolar. The main difference between polar and nonpolar dielectrics is that the atoms or molecules of polar dielectrics have an asymmetric shape, whereas the atoms or molecules of nonpolar dielectrics have a symmetric shape. Polar atoms or molecules have a permanent dipole moment, and thus, they behave like tiny electric dipoles. In the absence of an external electric field, the tiny dipoles are randomly arranged, and the net electric dipole moment of polar dielectrics is zero. Applying an external electric field forces the atoms or molecules within the polar dielectric medium to change their orientation and alignment to the electric field. The alignment of dipoles can be increased by increasing the external electric field and decreasing the temperature. On the contrary, nonpolar atoms and molecules do not have permanent dipole moment. When nonpolar atoms or molecules are subjected to an external electric field, the positive and negative charges are displaced in the opposite direction. This displacement continues until the external force is balanced by the restoring force due to the internal electric field. The internal electric field created due to the polarization of the dielectric is always opposite to the direction of the external electric field. Hence, the net electric field is reduced because of the polarization of the dielectric medium. However, when the external electric field is removed, the dipole moments of each nonpolar atom and molecule of the dielectric medium become zero.

In the case of a parallel plate capacitor, which consists of two parallel plates separated by a distance d, the electric field (E) is described by $E = V/d$, where V is a bias voltage applied to the capacitor. In vacuum, the charge (Q) on the plates is linearly proportional to the applied electric field and determined by Eq. (8):

$$Q = \varepsilon_0 E = \frac{\varepsilon_0 V}{d} \tag{8}$$

The ability of the capacitor to store charges is measured by its capacitance (C) and is defined by:

$$C = \frac{Q}{V} = \frac{\varepsilon_0}{d} \tag{9}$$

where ε_0 is the vacuum permittivity ($8.86 \times 10^{-12}\ C^2N^{-1}\ m^{-2}$), d is the distance between the plates, Q is the accumulated charge, and V is the potential difference between the plates. The polarization of a dielectric in a capacitor increases the capacitance by a factor equal to the relative permittivity ε_r of the material (also referred to as the dielectric constant k). Accordingly,

$$C = \varepsilon_0 \varepsilon_r \frac{A}{d} \tag{10}$$

where ε_0 is the vacuum permittivity ($8.86 \times 10^{-12}\ C^2N^{-1}\ m^{-2}$), ε_r is the dielectric permittivity, A is the plate overlap area, and d is the distance between the two plates.

As shown in Eq. (10), the capacitance varies directly with ε_r (k) and inversely with d. Consequently, in order to increase the capacitance of a parallel plate capacitor, two approaches are usually considered: the increase of C can be accomplished, first, by decreasing the dielectric thickness (d) and, second, by increasing the dielectric constant (k). However, conventional dielectric materials such as silicon dioxide or silicon nitride, which have been used abundantly in diverse applications throughout electronic devices, have reached their fundamental material limits, and decreasing their thickness below 2 nm is extremely challenging because it results in significantly increased leakage currents, which strongly affects the transistor operation reliability and its performance [18]. As a result, increasing C by employing high dielectric constant (high-k) materials using existing or novel high-k materials appears to be much more viable option. However, the development of new dielectric materials that possess high k and simultaneously show low leakage currents and high dielectric breakdown strength is not easy. Despite the obvious advantages of these materials in capacitor and OFET applications, they also possess serious drawbacks such as highly polar surfaces, which result in high charge trap densities particularly at the semiconductor/dielectric interface and polarization effects in the bulk, which lead to instability of the transistor threshold voltage and appearance of the source-drain current hysteresis [19]. In this light, many attempts to develop new high-k materials that can be inexpensively processed using novel, low cost deposition techniques have recently been carried out. Lately, a wide range of novel high dielectric constant materials ranging from high-k organic/inorganic nanocomposites, through multilayer high-k/low-k dielectric stacks to ultra-thin anodized oxides has been reported as promising alternatives to the conventional high-k insulators [20, 21]. In general, the dielectric materials used in OFETs are usually divided into four main categories, namely, inorganic [22, 23], organic [24, 25], electrolyte [26, 27], and hybrid dielectrics [28, 29].

3.2 Dielectrics in electronic devices

Dielectrics in electronic devices are usually utilized as insulators between conduction layers. A very important physical property of each dielectric is its energy gap (E_g). A large E_g is favorable because it requires electrons to acquire tremendous energy for excitation and transfer from the valence band to the conduction band. Usually, high-k dielectric materials have smaller E_g than SiO_2. In regard to the gate leakage current, small energy gaps may display a higher probability of direct tunneling across the dielectric by Schottky emission and/or Poole-Frenkel effect [30]. The relation between the energy gap and the dielectric constant of the most common inorganic high-k materials is illustrated in **Figure 4**. **Table 1** summarizes their most important electrical and structural properties. The major applications of high-k dielectrics are in capacitors [31], transistors [32], and memory devices such as dynamic random-access memory (DRAM) [33] and resistive memories (memristors) [34].

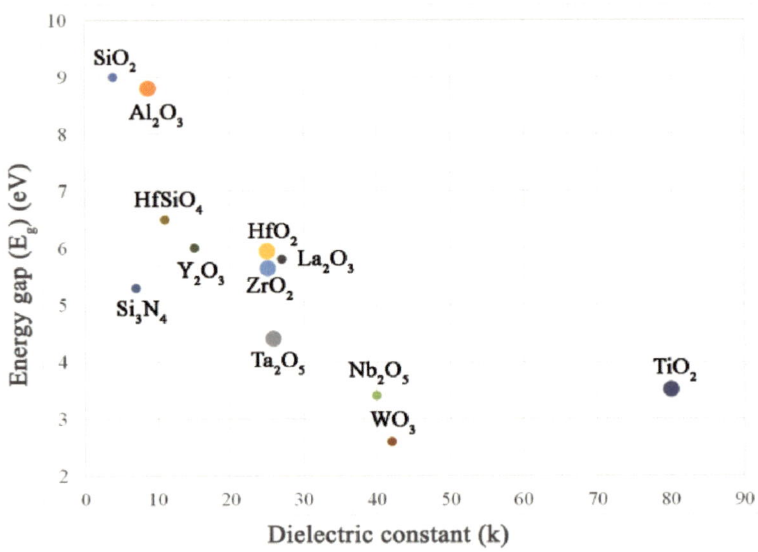

Figure 4.
Dielectric constant (k) vs. energy gap (E_g) for the most common inorganic dielectric materials [35].

Material	Dielectric constant (k)	Energy gap (eV)	Crystal structure
SiO_2	3.9	9	Amorphous
Al_2O_3	9	8.8	Amorphous
Ta_2O_5	26	4.4	Amorphous
HfO_2	25	5.8	Monoclinic, tetragonal, cubic
ZrO_2	25	5.8	Monoclinic, tetragonal, cubic
Nb_2O_5	40	3.4	Amorphous
TiO_2	80	3.5	Tetragonal (rutile, anatase, brookite)
WO_3	42	2.6	Monoclinic, tetragonal, rhombic
La_2O_3	27	5.8	Hexagonal
$HfSiO_4$	11	6.5	Tetragonal
Si_3N_4	5–7.5	5.3	Hexagonal, tetragonal
Y_2O_3	15	6	Cubic

Table 1.
The most important electrical, physical, and structural properties of some high-k inorganic materials [35].

3.3 Dielectrics in low voltage OFETs

3.3.1 Organic dielectrics

Organic dielectrics applied in OFETs are typically thicker than 200 nm because they are routinely deposited from solution by spin-coating, drop casting, or ink-jet printing, which result in low dielectric capacitance and subsequently in high operating voltage OFETs ($V_{GS} > \pm 20$ V) [36, 37]. Therefore, different methods have been used to overcome this problem including decreasing the gate dielectric thickness by depositing ultra-thin insulator films ($d < 10$ nm) [38, 39], utilizing high-k organic insulator materials [40], or doing both at the same time [41]. In case of using high-k materials [36], several groups have successfully employed high dielectric constant organic insulators and significantly lowered the operation voltage of OFETs. **Table 2** shows few examples of recently reported low voltage OFETs using organic dielectrics.

Among them, Li et al. used a high-k relaxor ferroelectric polymer as the gate dielectric ($k \approx 60$) and reduced the transistor operating voltage to 3 V [42]. Although V_G was reduced to 3 V, it has been found that highly polar dielectric materials possess high surface energy, which leads to increased trapping and, as a consequence, significantly lower field-effect mobility. Also, it turned out that the fluorinated surfaces of such materials are often incompatible with solution-processed

Ref.	Dielectric	Method	V_G (V)	C_i (nF/cm^2)	Semiconductor	μ (cm^2/Vs)	I_{ON}/I_{OFF}	Year
[39]	PVP	Spin coating	−3	250	Pentacene	0.5	$\sim 10^5$	2006
[17]	Cross-linked PVA	Spin coating	−2	12.2	TIPS-pentacene	1	$\sim 10^4$	2012
[42]	P(VDF-TrFE-CFE)	Spin coating	−3	330	pBTTTC16	0.4 ± 0.2	$\sim 10^4$	2012
[43]	PVA	Spin coating	−3	27	rr-P3HT	0.1	$\sim 10^3$	2012

Table 2.
Examples of low voltage OFETs using organic dielectrics.

organic semiconductors (OSCs) [43]. Machado and Hümmelgen have shown that using high-k cross-linked poly(vinyl alcohol) (crPVA, $k \sim 6.2$) as the gate dielectric might result in well-performing poly(3-hexyltiophene) (P3HT) OFETs [43]. However, their devices operated with $V_{GS} > 5$ V, and they suffered from relatively high leakage currents ($I_{Leak} > 20$ nA at $V_{GS} = 5$ V) [44]. Apart from using high-k insulator materials, a lot of efforts have been devoted to develop ultra-thin dielectric films ($d < 10$ nm), which can be processed from solution. Both cross-linked polymers of minimum thickness (<50 nm) [44] and self-assembled monolayers (SAMs) (e.g. octadecylphosphonic acid, ODPA) have been explored as the potential gate dielectric candidates for OFETs [45, 46]. However, cross-linked polymers and SAMs have usually been used in the bottom-gate OFETs, as they are difficult to be deposited on top of organic semiconductors (cf. **Figure 2**). Also, the possibility of contamination of the active layer by cross-linking agents, which may contribute to increased leakage currents and electrical instability of the transistors, limits the use of ultra-thin cross-linked polymer insulators to bottom-gate OFETs. In addition, it turns out that ultra-thin dielectrics are not fully compatible with low cost, high throughput printing techniques, and it is very challenging to process them reliably over large-area flexible substrates [47].

A promising way forward to address this problem and realize low voltage OFETs with low capacitance dielectrics is to use a material blend consisting of a small molecule organic semiconductor and an insulating polymer as the active layer. Using this approach, Feng et al. reported low voltage ($V_{GS} < 2$ V), solution-processed organic FETs with gate capacitance as small as 12.2 nF/cm^2 (cf. **Table 2**). This was achieved by employing a bottom-gate bottom-contact OFET architecture and using 6,13-bis(triisopropylsilylethynyl)-pentacene blended with polystyrene and UV cross-linked polyvinyl alcohol (PVA) as the active and the gate dielectric layers, respectively. It has been claimed that the low subthreshold swing value ($SS \sim 100$ mV/dec) was achieved due to a significant decrease in the effective channel capacitance described by Eq. (3) and very smooth PVA surface with a root-mean-square (RMS) roughness 0.3 nm, which contributed to the exceptionally low interface trap density.

In summary, to realize low voltage operated OFETs, low values of threshold voltage and subthreshold swing are required. Alternatively, one can achieve low voltage operated OFETs reducing the number of traps optimizing the dielectric/semiconductor interface. However, both approaches are not trivial, and more materials and device research are needed to find the optimal solutions for the intended applications.

3.3.2 Inorganic dielectrics

Today silicon is the most used material in the electronic industry. Si can be reacted with oxygen to form excellent dielectrics [1]. However, SiO_2 has relatively low dielectric constant ($k = 3.9$), and therefore, it is rather problematic to realize low voltage ($1 \leq V_G \leq 3$) and ultra-low voltage ($V_G \leq 1$) OFETs. Decreasing the thickness of SiO_2 to achieve the required capacitances is extremely difficult because of the charge tunneling effect that significantly increases the gate leakage current when SiO_2 is thinner than 2 nm. In this case, several alternative metal oxide dielectrics (e.g. Al_2O_3, HfO_2, Ta_2O_5, ZrO_2, TiO_2, Y_2O_3, CeO_2, etc.) have been investigated to be used as a gate insulator in OFETs [48]. Herein, we focus on the two most promising metal oxide dielectrics for OFETs, namely, Al_2O_3 and Ta_2O_5.

3.3.2.1 Aluminum oxide (Al_2O_3)

Aluminum oxide is an inert, water insoluble metal oxide, which due to large energy gap ($E_g \sim 8.8$ eV), high dielectric constant ($k \sim 9$), and the low cost is abundantly used in the electronic industry as an insulator [13]. The competent insulating behavior makes Al_2O_3 particularly suitable for low voltage OFETs. Thanks to its amorphous crystal structure, it can be deposited using a wide range of deposition techniques including r.f. magnetron sputtering, plasma-assisted oxidation, sol-gel, anodization, and so on. Also, the availability of aluminum in the form of plastic aluminized foils makes it a potential material of choice for the next generation of smart electronic goods. Al_2O_3 can be very thin ($d \leq 3$ nm) and still maintain excellent insulating properties. As such, it has received a lot of attention from different research groups working on low voltage inorganic TFTs and OFETs in recent years. For example, in 2011, Avis et al. proposed a 70-nm thick sol-gel AlO_x as the gate dielectric for 5 V zinc-tin-oxide (ZTO) thin-film transistors [49]. In the same year, Lan et al. proposed a 140 nm anodic Al_xO_y for using in indium oxide (In_2O_3) and indium-gallium-zinc-oxide (IGZO) TFTs [50]. Although the reported transistors operated at 6 V, their application may be somewhat limited due to the high temperature processing ($T \sim 300°C$). Even though the abovementioned TFTs are incompatible with most of flexible plastic substrates, they can still be used in printed electronics but on high temperature plastic films or rigid substrates (e.g. glass). In parallel research, Chen et al. proposed high performance, low voltage ZnO TFTs employing 100 nm Al_2O_3 deposited by DC magnetron sputtering as the gate dielectric [51]. The minimum operating voltage of the proposed devices was 4 V, but due to the thick Al_2O_3, the transistors could not be operated with lower V_{GS}. In 2017, Cai et al. reported 1 V IGZO TFTs that employed a 3 nm thick solution processed anodic Al_2O_3 [52]. The demonstrated devices operated at 1 V, had on/off current ratios larger than 10^5, displayed field-effect mobilities of around 5.4 cm^2/V·s, and possessed subthreshold swing of 68 mV/dec, which is close to the theoretical limit of *SS* at 300 K. In 2018, Ma et al. proposed low voltage IGZO TFTs using a 5 nm Al_2O_3 dielectric that resulted in transistors operating at 0.6 V [53].

In addition, there were few attempts to use pristine aluminum oxide as the gate insulator in OFETs. For example, Shang et al. proposed low threshold voltage pentacene OFETs and circuits [54]. The demonstrated devices possessed field-effect mobility 0.16 cm^2/Vs, I_{ON}/I_{OFF} current ratio about 10^5, threshold voltage 0.3 V, and subthreshold swing 0.6 V/decade. The low voltage device was achieved by growing the oxide layer using atomic layer deposition (ALD) technique. ALD provides high quality, pinhole free oxide layers and is typically used for high performance TFTs and FETs. However, this material deposition method requires very expensive equipment, and the materials have to be synthesized in high vacuum, which may not meet the demands of low cost, room temperature, large-area manufacturing of electronics. Sun et al. reported 3 V pentacene OFETs using 50 nm thick solution deposited Al_2O_3 [55]. The demonstrated devices possessed field-effect mobility near 3 cm^2/Vs, I_{ON}/I_{OFF} current ratio about 10^6, threshold voltage -0.9 V, and subthreshold swing 107 mV/decade. However, it appears that the best performing OFETs have been obtained with SAM-modified Al_2O_3 where self-assembled monolayers are used as a buffer between the organic semiconductor and the aluminum oxide [56, 57].

3.3.2.2 Tantalum pentoxide (Ta_2O_5)

Tantalum pentoxide (Ta_2O_5) is a highly promising dielectric material because it has high transparency, high melting point (1785°C), high dielectric constant and shows good thermal and chemical stabilities. As such, it has been used in a wide range of electronic applications such as in dynamic random-access memory (DRAM), metal-insulator-metal (MIM) capacitors, memory resistors (memristors), and recently in organic and inorganic TFTs [58]. The dielectric constant of Ta_2O_5

Ref.	Dielectric	Method	d (nm)	V_G (V)	C_i (nF/ cm^2)	Semi-conductor	μ (cm^2/ Vs)	I_{ON}/ I_{OFF}	Year
[49]	Al_2O_3	Sol–gel	70	3	80	ZTO	33	$\sim10^8$	2011
[50]	Al_2O_3	Anodization	140	4	54	IGZO	21.6	$\sim10^8$	2011
[51]	Al_2O_3	D.C. sputtering	100	4	117	ZnO	27	$\sim10^6$	2012
[52]	Al_2O_3	Anodization	3	1	1000	IGZO	5.4	$\sim10^5$	2017
[53]	Al_2O_3	ALD	5	0.6	720	IGZO	3.8	$\sim10^6$	2018
[54]	Al_2O_3	ALD	30	−3	165	Pentacene	0.16	$\sim10^5$	2011
[55]	Al_2O_3	Spin coating	50	−3	125	Pentacene	2.7	$\sim10^6$	2016
[60]	Ta_2O_5	R.f. sputtering	130	−3	163	Pentacene	0.8	66	2004
[61]	Ta_2O_5	e-beam	200	3	89	a-IGZO	61.5	$\sim10^5$	2010
[62]	Ta_2O_5	e-beam	100	−2	185	P3HT	0.02	$\sim10^5$	2002

Table 3.
Examples of low voltage TFTs and OFETs using inorganic dielectrics.

depends on its thickness and deposition technique and varies from ~35 in the bulk to ~25 in a thin film [59]. This is at least two to six times larger than the dielectric constant of Al_2O_3 (k = 9) and SiO_2 (k = 3.9), respectively. As a result, Ta_2O_5 appears to be a good candidate as the gate dielectric in OFETs. Ta_2O_5 can be grown using different methods, namely, thermal oxidation, plasma-assisted oxidation, r.f. sputtering, atomic layer deposition, and anodization [60]. However, its dielectric properties can significantly differ depending on which of the growth methods is used. Recently, Chiu et al. have used a 200 nm e-beam deposited Ta_2O_5 for amorphous indium-gallium-zinc oxide (a-IGZO) TFTs, which operated at 3 V [61]. Lately, Bartic et al. have proposed 3 V bottom-gate bottom-contact and top-gate bottom-contact P3HT OFETs employing a 100 nm film of Ta_2O_5 deposited by e-beam evaporation as the gate dielectric [62]. Although both inorganic and organic transistors have yielded high performance, e-beam evaporation is a relatively expensive deposition technique, and it is not compatible with the idea of low cost, large-area electronics. In addition, thick Ta_2O_5 films employed in the aforementioned TFTs made it impossible to operate the transistors at or below 1 V, which hinders their use in special applications such as portable, ultra-low power electronics, or aqueous sensors. **Table 3** summarizes all device parameters of the above-discussed TFTs and OFETs.

3.3.3 Organic-inorganic bilayer dielectrics

3.3.3.1 Self-assembled monolayers (SAMs)

Self-assembled monolayers (SAMs) are ordered, two-dimensional organic molecular assemblies formed spontaneously by chemical absorption of an amphiphilic surfactant on a variety of substrates. In particular, silane SAMs are long-chain hydrocarbon molecules, which form an ordered supramolecular structure on solid surfaces after absorption. As such, they have been vastly used for surface modification and capping. One highly promising way to suppress the insulator surface charge traps in OFETs is to treat the transistor dielectric surface with hydrophobic SAMs, such as hexamethyldisilane (HMDS) [63], octyltrichlorosilane (OTS),

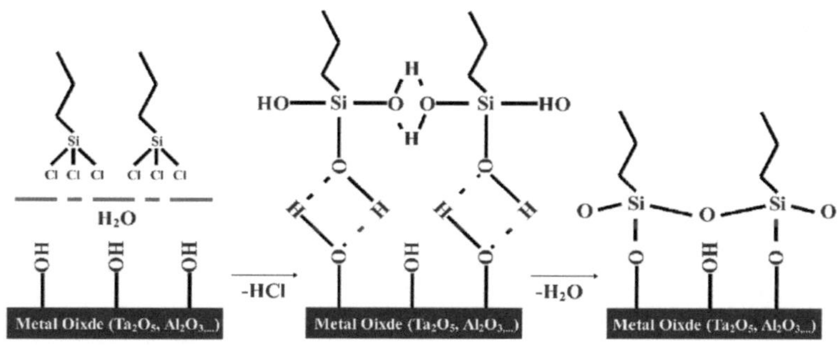

Figure 5.
Schematic representation of the silanization reaction.

or *n*-octadecyltrichlorosilane (ODTS). As reported in [64, 65], carrier mobility is significantly improved with increasing the SAM alkyl chain length, i.e., HMDS < OTS < ODTS, in comparison with the untreated surface [66]. Also, having a longer alkyl chain SAM improves adhesion and hydrophobicity of the modified surface [67].

n-octadecyltrichlorosilane (ODTS) is a self-assembled monolayer that has previously been shown to have good compatibility with SiO_2 and metal oxide dielectrics. Nowadays, it is typically used as a passivation layer for metal oxides providing capping of polar surfaces or as a hydrophobic coating layer preventing electrical instability of organic semiconductors and OFETs [3]. Essentially, ODTS appears to be one of the most used SAMs in organic FETs. It has been reported that ODTS significantly improves dielectric/semiconductor interface by passivating the metal oxide dielectric surface that leads to the reduction of charge carrier traps and, in consequence, to higher charge carrier mobility [66]. During silanization, ODTS molecules are attached to the dielectric surface through the chemical reaction of –SiCl with –OH groups on the metal oxide surface. This results in –Si–O–M structures. The other two –SiCl bonds of the ODTS molecule react with proximate OTS molecules, which form a cross-linked monolayer (**Figure 5**).

3.3.3.2 Organic-inorganic hybrid dielectrics

To achieve the best of both worlds, several research groups have been researching the organic-inorganic bilayer and multilayer dielectrics (also known as hybrid and high-k/low-k dielectric) for low voltage OFETs. Liu et al. reported polymer field-effect transistors utilizing two diketopyrrolopyrrole (DPP)-based copolymers (i.e. PDQT and PDVT-10) as the semiconductor and OTS-modified poly(vinyl alcohol) (PVA) as the gate dielectric [68]. Their devices operated at around 3 V, and it was claimed that the OTS modification of PVA enhanced carrier mobility, lowered the leakage current, resulted in less hysteresis, and generally led to better performing devices than OFETs with untreated PVA. Urasinska-Wojcik et al. fabricated 1 V organic FETs using a mixed SAM/Al_2O_3 bilayer as the gate dielectric and poly(3,6-di (2-thien-5-yl)-2,5-di (2-octyldodecyl)-pyrrolo([3,4-c]pyrrole-1,4-dione) thieno [3,2-b] thiophene) (DPPDTT) as the organic semiconducting layer [57]. As reported, the self-assembled monolayer surface modification made Al_2O_3 surface smoother, and it was concluded that the SAM passivation helped tuning the threshold voltage and improved field-effect mobility of the proposed OFETs when compared with untreated devices. Mohammadian et al. proposed 1 V OFETs gated by ODTS-treated Ta_2O_5 and DPPDTT as the organic semiconductor [69]. The proposed transistors operated with the field-effect carrier mobility around 0.2 $cm^2V^{-1}\ s^{-1}$, threshold

Ref.	Dielectric	Method	d (nm)	V_G (V)	C_i (nF/cm^2)	OSC	μ (cm^2/Vs)	I_{ON}/I_{OFF}	Year
[68]	PVA/OTS	Spin-coating	230 + OTS thickness	3	28	PDVT-10	11	~10^4	2014
[57]	Al_2O_3/OTS	Anodization	4 + ODTS thickness	1	340	PDPP2TTT	0.1	~10^3	2015
[69]	Ta_2O_5/ODTS	Anodization	4 + ODTS thickness	1	670	DPPDTT	0.2	5×10^3	2019

Table 4.
Parameters of the previously demonstrated low voltage OFETs using organic/inorganic bilayer dielectrics.

voltage −0.55 V, subthreshold swing 120 mV/dec, and current on/off ratio in excess of 5×10^3. The ODTS surface treatment used in the reported OFETs did not only make the surface smoother and improve the charge carrier mobility but also was used as a support of the main Ta_2O_5 dielectric. It was found that adding the extra insulator layer to the high-k dielectric increased its overall thickness and therefore decreased the gate capacitance, but because of low threshold voltage (V_T = −0.55 V), 1 V OFET operation was still possible. The same approach toward the gate dielectric engineering for low voltage OFETs was reported in [70, 71]. **Table 4** summarizes the key parameters of all above-discussed OFETs.

3.4 High-k metal oxide deposition techniques

3.4.1 Radio frequency (r.f.) magnetron sputtering

Radio frequency (r.f.) magnetron sputtering is a thin-film vapor deposition (PVD) technique. The process begins when a voltage is applied to a target material in the presence of argon gas. In such an instance, plasma is created in the surrounding of the target and ionized argon gas molecules start to bombard the target atoms. This bombardment leads the atoms to be sputtered off into the plasma. Then, these vaporized atoms are deposited when they condense as a thin film on the substrate. In order to properly deposit the sputtered materials, several process parameters should carefully be considered. First, the distance between the samples and the target should be optimized. Second, the chamber pressure should carefully be controlled to get the best quality of the deposited films. Last but not least the applied sputtering power should not exceed the maximum value for a given material because higher applied power could result in the target damage and poor quality of the deposited films. Usually, for Ta deposition r.f. magnetron sputtering is performed in the optimum pressure $P = 5 \times 10^{-3}$ mBar, samples are 10 cm apart from the target and power does not exceed 70 W. **Figure 6** demonstrates the schematic of the r.f. magnetron sputtering deposition process.

3.4.2 Atomic layer deposition (ALD)

Atomic layer deposition (ALD) is a popular material deposition method, which is a subclass of chemical vapor deposition (CVD) technique. ALD is a high yield process delivering a highly conformal, pinhole free oxide layers at a relatively low temperature. In this process, two chemicals react with each other, and the oxide is achieved by repeating sequential, self-limiting surface reactions where precursors are separately deposited onto the substrate [59]. One ALD advantage in comparison with other vacuum deposition methods is that the oxide layer grows per cycle (GPC) allowing to have a sub-nanometer control over the deposited layer. ALD has two main drawbacks. First, the temperature of the oxidation process is relatively

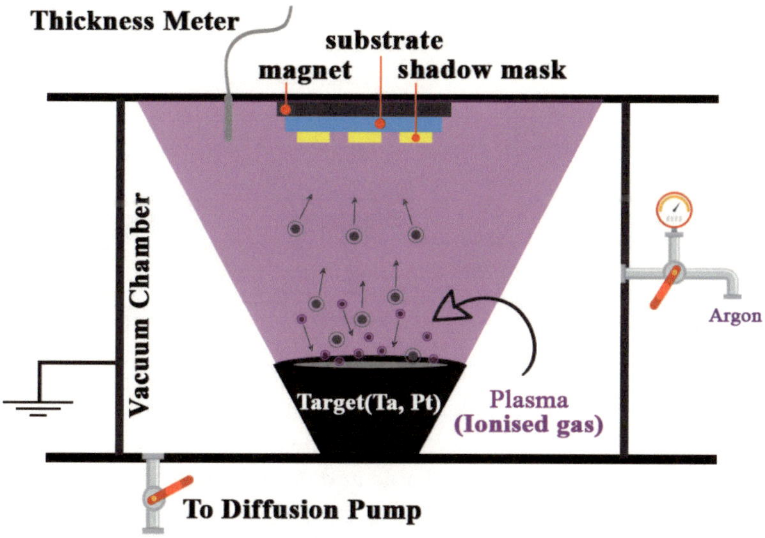

Figure 6.
Schematic of the r.f. magnetron sputtering process.

high ($T \sim 300°C$), and thus it cannot be used for majority of plastic substrates. Second, a number of research groups reported several percentages of carbon contamination in the deposited oxide layers [72].

3.4.3 Plasma-assisted oxidation

Plasma-assisted oxidation is a widespread oxidation technique, which relies on oxidation of materials using highly reactive oxygen species. In general, oxidation of metals up to 2 nm is a straightforward process, but deeper oxidations are difficult to produce because the initial layer shields further oxidation. The description of the complete plasma oxidation process is complex and consists of several parts: volume processes in chemically active plasma of oxygen or its mixtures with inert gases, transport of particles through the transient region between undisturbed plasma and metal sample immersed into plasma, processes on the surface of the sample, and the transport of both oxygen and metal ions through the growing oxide layer. Importantly, oxygen plasma is also widely used for cleaning, etching, and removing unwanted organic residues from surfaces [73]. Hsiao et al. have shown that high quality Al_2O_3 and Ta_2O_5 can be successfully grown by plasma-assisted oxidation in a controlled environment [72].

3.4.4 Solution-based techniques: spin-coating

Spin coating is a common method to produce thin, uniform polymer films on flat substrates. In the spin-coating process (**Figure 7**), the solution is first deposited on the substrate, and then it is accelerated rapidly to a desirable spin speed. This method is normally used for deposition of polymer films with thicknesses, which range from few nanometers to several micrometers. Depending on the spin speed, spin acceleration, and viscosity (concentration) of the solution, the thickness of the deposited layers can be precisely controlled [74].

3.4.5 Other solution-based techniques: anodization

The material discussed in this section is tantalum oxide (Ta_2O_5), which is generally used as an insulator in electronic devices such as capacitors, memristors,

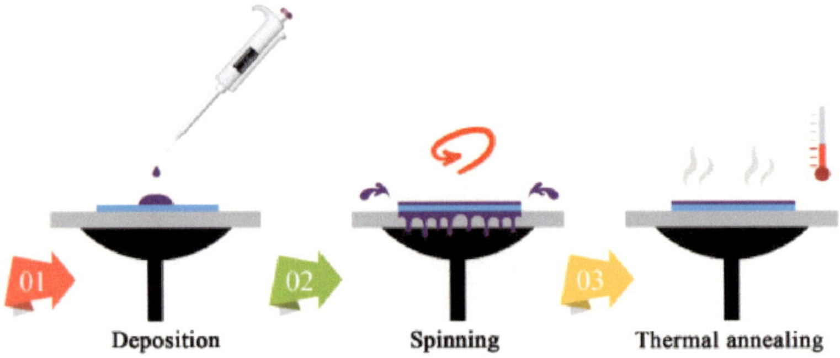

Figure 7.
Schematic of the spin-coating process.

TFTs, and OFETs. One of the most reliable and straightforward ways to form tantalum oxide is electrochemical oxidation (so-called anodization). Several metals such as Al, Ti, and Ta have very high chemical affinity with oxygen. Therefore, under ambient conditions, they will rapidly react with this gas forming a "native oxide" on their surfaces. The "air-formed" oxide film protects the metals from further oxidation (i.e. oxidation of the metal bulk) but is extremely thin—its thickness varies from a few angstroms to circa 2 nm, and it is very often not homogenous in thickness and can contain numerous defects and flaws. For example, the native oxide layer of tantalum has been reported to be usually around 3–4 nm [75]. As a result, the native oxide cannot be used as a protective film for preventing corrosion or as an insulator in capacitors and transistors.

Anodization is an electrochemical process, which allows improving this natural oxide film and produces stable oxide films with negligible reactivity. **Figure 8** shows a typical anodization bath to perform the oxidation. During the process, the metal to be oxidized is made the anode. An electrolytic cell is filled with an electrolyte. It has been shown previously that the nature of the electrolyte determines the type of anodic oxide film. The electric circuit is completed with a counter electrode which is made of a chemically inert metal (e.g. Au, Pt) or alloy (e.g. stainless steel). In **Figure 8**, an Au plate is shown as the cathode.

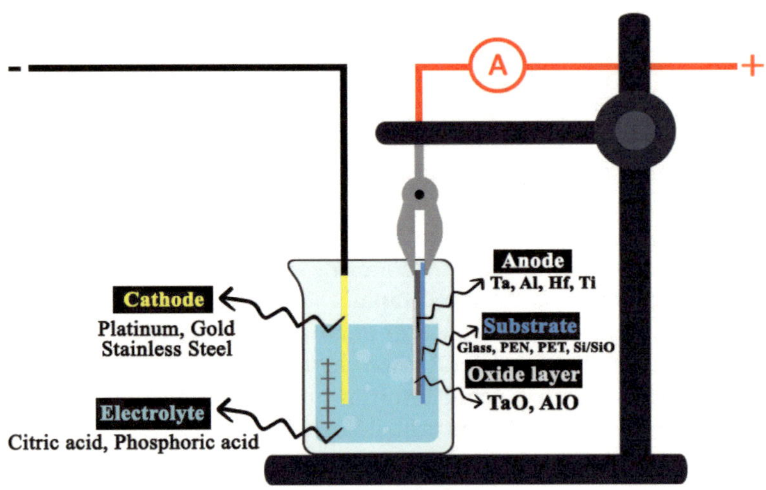

Figure 8.
Schematic of the anodization process.

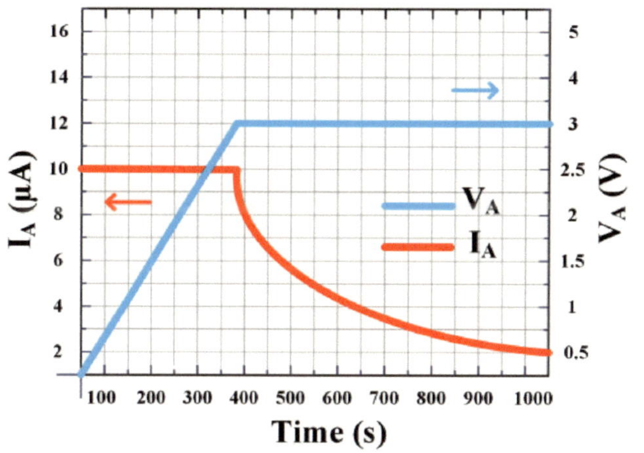

Figure 9.
Anodization voltage (V_A) and anodization current (I_A) vs. anodization time (t).

Figure 9 illustrates the electric drive conditions for anodization in constant current mode (I_A = const.). In the constant current mode, the thickness of an oxide depends only on the cell voltage, which is allowed to rise to the required value, i.e. the thickness d of the resulting films can be very precisely controlled via anodization voltage V_A because $d = c \cdot V_A$, where c is the anodization ratio describing the thickness of the formed film per applied volt (Å/V). Importantly, c is related to the electric breakdown field E_B via $E_B \approx c^{-1}$.

The anodization current I_A is kept constant via the anodization voltage V_A compensation until the desired voltage is achieved and then decreases to very low values. The "leakage current" flowing under constant voltage conditions is electronic, but if the voltage is increased, then ionic current begins to flow again with further film formation until a new equilibrium is established. However, it is not possible to increase the voltage to a very high value. The upper limit on voltage lies between 500 and 700 V due to breakdown and arcing in the barrier layer. The most interesting fact is that the thickness of the barrier type film is not affected by electrolyzing time, surface roughness, and temperature of the electrolyte. In fact, the formed oxide film will exactly follow or slightly smooth out the initial surface topography of the anodized metal.

The anodization ratio of tantalum anodized in 1 mM citric acid (CA) has been reported in the literature [74–76] and usually is between 1.8 and 2.2 nm/V. **Table 5** compares the information relating to the anodization of metals and in particular anodization of tantalum in the recent works [68, 78–80].

Ref.	Metal oxide	Anodization ratio (nm/V)	Forming electrolyte
[74]	Ta_2O_5	1.8	0.1 M H_3PO_4
[75]	Ta_2O_5	2.2	0.01 M CA
[76]	Ta_2O_5	2.0	0.01 M CA
[77]	TiO_2	1.5	0.001 M CA
[66]	Al_2O_3	1.3	0.001 M CA

CA—citric acid.

Table 5.
A summary of anodization ratios and forming electrolytes relating to the anodization of Al_2O_3, TiO_2 and Ta_2O_5.

4. Conclusions

In this chapter, the most promising strategies toward lowering the operational voltage of organic FETs have been reviewed and discussed. This includes reducing the transistor threshold voltage and subthreshold swing. Apart from the semiconductor/insulator interface engineering that is not always straightforward, one either can employ high-k dielectric materials, reduce their thickness, or do both at the same time. The best performing dielectric materials in OFETs appear to be metal oxides. They intrinsically possess high dielectric constants and display low leakage currents. Also, they can be made ultra-thin ($d \leq 3$ nm), and when deposited on plastic films, they are flexible and robust. However, depending on the intended applications, one can also use pristine organic, organic-inorganic hybrid, or high-k/low-k multilayer dielectrics. Anodic oxidation is a very promising technique, which can considerably lower manufacturing costs of high-k materials and realize inexpensive low voltage OFETs and OFET-based circuits. It is a cheap, solution-based deposition process that can be performed under ambient conditions. Since the anodization is a self-limiting and self-healing process, it can give pinhole-free, homogenous oxide layers that can be grown in ambient atmosphere at room temperature. As such, anodization has a high potential to be used in manufacturing of future OFET-based electronic devices and circuits.

Author details

Navid Mohammadian* and Leszek A. Majewski*
Department of Electrical and Electronic Engineering, The University of Manchester, Manchester, United Kingdom

*Address all correspondence to: navid.mohammadian@manchester.ac.uk and leszek.majewski@manchester.ac.uk

References

[1] Sze SM, Ng KK. Physics of Semiconductor Devices Physics of Semiconductor Devices. US: Wiley-Interscience; 1995. p. 793. DOI: 10.1002/0470068329

[2] Horowitz G. Organic field-effect transistors. Advanced Materials. 1998; **10**(5):365-377

[3] Gomes HL. Organic Field-Effect Transistors. Organic and Printed Electronics Fundamentals and Applications. Boca Raton, FL: CRC Press; 2016. pp. 147-197

[4] Kymissis I. Organic Field Effect Transistors: Theory, Fabrication and Characterization. US: Springer Science & Business Media; 2008

[5] Horowitz G. The organic transistor: state-of-the-art and outlook. The European Physical Journal - Applied Physics. EDP Sciences. 2011;**53**(3): 33602. DOI: 10.1051/epjap/2011100055

[6] Wöll C. Physical and Chemical Aspects of Organic Electronics: From Fundamentals to Functioning Devices. Germany: John Wiley & Sons; 2009

[7] Su Y, Wang C, Xie W, Xie F, Chen J, Zhao N, et al. Low-voltage organic field-effect transistors (OFETs) with solution-processed metal-oxide as gate dielectric. ACS Applied Materials & Interfaces. 2011;**3**(12):4662-4667

[8] Yang F, Sun L, Han J, Li B, Yu X, Zhang X, et al. Low-voltage organic single-crystal field-effect transistor with steep subthreshold slope. ACS Applied Materials & Interfaces. 2018;**10**(31): 25871-25877

[9] Kunii M, Iino H, Hanna JI. Solution-processed, low-voltage polycrystalline organic field-effect transistor fabricated using highly ordered liquid crystal with low-k gate dielectric. IEEE Electron Device Letters. 2016;**37**(4):486-488

[10] Street R. Technology and Applications of Amorphous Silicon. Vol. 37. Germany: Springer Science & Business Media; 1999

[11] Elkington D, Cooling N, Belcher W, Dastoor P, Zhou X. Organic thin-film transistor (OTFT)-based sensors. Electronics [Internet]. 2014;**3**(2): 234-254 [cited: 25 January 2020]

[12] Liao F, Chen C, Subramanian V. Organic TFTs as gas sensors for electronic nose applications. Sensors and Actuators B: Chemical. 2005;**107**(2): 849-855

[13] Ortiz RP, Facchetti A, Marks TJ. High-k organic, inorganic, and hybrid dielectrics for low-voltage organic field-effect transistors. Chemical Reviews. 2010;**110**(1):205-239. DOI: 10.1021/cr9001275

[14] Yokota T, Sekitani T, Kato Y, Kuribara K, Zschieschang U, Klauk H, et al. Low-voltage organic transistor with subfemtoliter inkjet source–drain contacts. MRS Communications. 2011; **1**(1):3-6

[15] Feng L, Tang W, Zhao J, Yang R, Hu W, Li Q, et al. Unencapsulated air-stable organic field effect transistor by all solution processes for low power vapor sensing. Scientific Reports. 2016; **6**:20671

[16] Park YD, Kim DH, Jang Y, Hwang M, Lim JA, Cho K. Low-voltage polymer thin-film transistors with a self-assembled monolayer as the gate dielectric. Applied Physics Letters [Internet]. 2005;**87**(24):243509. DOI: 10.1063/1.2143113

[17] Feng L, Tang W, Xu X, Cui Q, Guo X. Ultralow-voltage solution-processed organic transistors with small gate dielectric capacitance. IEEE Electron Device Letters. 2013;**34**(1): 129-131. DOI: 10.1109/LED.2012.2227236

[18] Wilk GD, Wallace RM, Anthony J. High-κ gate dielectrics: Current status and materials properties considerations. Journal of Applied Physics. 2001; **89**(10):5243-5275

[19] Smith J, Hamilton R, McCulloch I, Heeney M, Anthony JE, Bradley DDC, et al. High mobility p-channel organic field effect transistors on flexible substrates using a polymer-small molecule blend. Synthetic Metals. 2009; **159**(21–22):2365-2367

[20] Tang W, Zhao J, Huang Y, Ding L, Li Q, Li J, et al. Bias stress stability improvement in solution-processed low-voltage organic field-effect transistors using relaxor ferroelectric polymer gate dielectric. IEEE Electron Device Letters. 2017;**38**(6):748-751

[21] Shin E, Cho HJ, Jung S, Yang C, Noh Y. A high-k fluorinated P (VDF-TrFE)-g-PMMA gate dielectric for high-performance flexible field-effect transistors. Advanced Functional Materials. 2018;**28**(4):1704780

[22] Ma P, Sun J, Liang G, Li Y, Xin Q, Li Y, et al. Half-volt operation of IGZO thin-film transistors enabled by ultrathin HfO_2 gate dielectric. Applied Physics Letters. 2018;**113**(6): 63501

[23] Xia G, Wang S. Rapid and facile low-temperature solution production of ZrO_2 films as high-k dielectrics for flexible low-voltage thin-film transistors. Ceramics International. 2019;**45**(13):16482-16488

[24] Wang Y, Huang X, Li T, Wang Z, Li L, Guo X, et al. Novel crosslinkable high-k copolymer dielectrics for high-energy-density capacitors and organic field-effect transistor applications. Journal of Materials Chemistry A. 2017; **5**(39):20737-20746

[25] Spanu A, Pinna L, Viola F, Seminara L, Valle M, Bonfiglio A, et al. A high-sensitivity tactile sensor based on piezoelectric polymer PVDF coupled to an ultra-low voltage organic transistor. Organic Electronics. 2016;**36**: 57-60

[26] Na Y, Kim FS. Nanodroplet-embedded semiconducting polymer layers for electrochemically stable and high-conductance organic electrolyte-gated transistors. Chemistry of Materials. 2019;**31**(13):4759-4768

[27] Zhang Q, Leonardi F, Pfattner R, Mas-Torrent M. A solid-state aqueous electrolyte-gated field-effect transistor as a low-voltage operation pressure-sensitive platform. Advanced Materials Interfaces. 2019;**6**(16):1900719

[28] Kim D-K, Choi J-H. Low-voltage organic light-emitting field-effect transistors using n-dodecylphosphonic acid-passivated HfO_x dielectrics. Organic Electronics. 2017;**51**:287-294

[29] Lee S-H, Kim D-Y, Noh Y-Y. Solution-processed polymer-sorted semiconducting carbon nanotube network transistors with low-k/high-k bilayer polymer dielectrics. Applied Physics Letters. 2017;**111**(12):123103

[30] Qin HL, Troadec C, Goh KEJ, Kakushima K, Iwai H, Bosman M, et al. Electronic properties of ultrathin high-κ dielectrics studied by ballistic electron emission microscopy. Journal of Vacuum Science & Technology, B: Nanotechnology & Microelectronics: Materials, Processing, Measurement, & Phenomena. 2011;**29**(5):52201

[31] Ding S-J, Hu H, Lim HF, Kim SJ, Yu XF, Zhu C, et al. High-performance

MIM capacitor using ALD high-k HfO_2-Al_2O_3 laminate dielectrics. IEEE Electron Device Letters. 2003;**24**(12):730-732

[32] Choi J, Joo M, Seong H, Pak K, Park H, Park CW, et al. Flexible, low-power thin-film transistors made of vapor-phase synthesized high-k, ultrathin polymer gate dielectrics. ACS Applied Materials & Interfaces. 2017;**9**(24):20808-20817

[33] Asano I, Nakamura Y, Hiratani M, Nabatame T, Iijima S, Saeki T, et al. Development of MIM/Ta_2O_5 capacitor process for 0.10-μm DRAM. Electronics and Communications in Japan (Part II Electronics). 2004;**87**(2):26-36

[34] Yoon JH, Song SJ, Yoo I, Seok JY, Yoon KJ, Kwon DE, et al. Highly uniform, electroforming-free, and self-rectifying resistive memory in the Pt/Ta_2O_5/HfO_{2-x}/TiN structure. Advanced Functional Materials. 2014;**24**(32):5086-5095

[35] Sunami H. Dimension increase in metal-oxide-semiconductor memories and transistors. In: Advanced in Solid State Circuit Technologies. Rijeka, Croatia: InTech; 2010. pp. 307-332

[36] Sheraw CD, Zhou L, Huang JR, Gundlach DJ, Jackson TN, Kane MG, et al. Organic thin-film transistor-driven polymer-dispersed liquid crystal displays on flexible polymeric substrates. Applied Physics Letters. 2002;**80**(6):1088-1090

[37] McDowell M, Hill IG, McDermott JE, Bernasek SL, Schwartz J. Improved organic thin-film transistor performance using novel self-assembled monolayers. Applied Physics Letters. 2006;**88**(7):73505

[38] Ma H, Acton O, Ting G, Ka JW, Yip H-L, Tucker N, et al. Low-voltage organic thin-film transistors with π-σ-phosphonic acid molecular dielectric monolayers. Applied Physics Letters. 2008;**92**(11):99

[39] Yang SY, Kim SH, Shin K, Jeon H, Park CE. Low-voltage pentacene field-effect transistors with ultrathin polymer gate dielectrics. Applied Physics Letters [Internet]. 2006;**88**(17):173507. DOI: 10.1063/1.2199592

[40] Zirkl M, Haase A, Fian A, Schön H, Sommer C, Jakopic G, et al. Low-voltage organic thin-film transistors with high-k nanocomposite gate dielectrics for flexible electronics and optothermal sensors. Advanced Materials. 2007;**19**(17):2241-2245

[41] Tang W, Li J, Zhao J, Zhang W, Yan F, Guo X. High-performance solution-processed low-voltage polymer thin-film transistors with low-k/high-k bilayer gate dielectric. IEEE Electron Device Letters. 2015;**36**(9):950-952. DOI: 10.1109/LED.2015.2462833

[42] Li J, Sun Z, Yan F. Solution processable low-voltage organic thin film transistors with high-k relaxor ferroelectric polymer as gate insulator. Advanced Materials [Internet]. 2012;**24**(1):88-93. DOI: 10.1002/adma.201103542

[43] Machado WS, Hummelgen IA. Low-voltage poly (3-hexylthiophene)/poly (vinyl alcohol) field-effect transistor and inverter. IEEE Transactions on Electron Devices. 2012;**59**(5):1529-1533

[44] Li Y, Wang H, Shi Z, Mei J, Wang X, Yan D, et al. Novel high-k polymers as dielectric layers for organic thin-film transistors. Polymer Chemistry. 2015;**6**(37):6651-6658

[45] Cheng X, Caironi M, Noh Y-Y, Wang J, Newman C, Yan H, et al. Air stable cross-linked cytop ultrathin gate dielectric for high yield low-voltage top-gate organic field-effect transistors.

Chemistry of Materials. 2010;**22**(4): 1559-1566

[46] Klauk H, Zschieschang U, Pflaum J, Halik M. Ultralow-power organic complementary circuits. Nature. 2007; **445**(7129):745-748

[47] Wöbkenberg PH, Ball J, Kooistra FB, Hummelen JC, de Leeuw DM, Bradley DDC, et al. Low-voltage organic transistors based on solution processed semiconductors and self-assembled monolayer gate dielectrics. Applied Physics Letters. 2008;**93**(1):242

[48] Majewski LA, Schroeder R, Grell M. Organic field-effect transistors with ultrathin gate insulator. Synthetic Metals. 2004;**144**(1):97-100

[49] Avis C, Jang J. High-performance solution processed oxide TFT with aluminum oxide gate dielectric fabricated by a sol–gel method. Journal of Materials Chemistry. 2011;**21**(29): 10649-10652

[50] Lan L, Peng J. High-performance indium–gallium–zinc oxide thin-film transistors based on anodic aluminum oxide. IEEE Transactions on Electron Devices. 2011;**58**(5):1452-1455

[51] Chen R, Zhou W, Zhang M, Kwok HS. High performance self-aligned top-gate ZnO thin film transistors using sputtered Al_2O_3 gate dielectric. Thin Solid Films. 2012; **520**(21):6681-6683

[52] Cai W, Zhang J, Wilson J, Song A. Low-voltage, full-swing InGaZnO-based inverters enabled by solution-processed, ultra-thin Al_xO_y. IEEE Electron Device Letters. 2019;**40**(8): 1285-1288

[53] Ma P, Du L, Wang Y, Jiang R, Xin Q, Li Y, et al. Low voltage operation of IGZO thin film transistors enabled by ultrathin Al_2O_3 gate dielectric. Applied Physics Letters. 2018;**112**(2):23501

[54] Shang L, Ji Z, Chen Y, Wang H, Liu X, Han M, et al. Low voltage organic devices and circuits with aluminum oxide thin film dielectric layer. Science China Technological Sciences [Internet]. 2011;**54**(1):95-98. DOI: 10.1007/s11431-010-4213-z

[55] Sun Q-J, Peng J, Chen W-H, She X-J, Liu J, Gao X, et al. Low-power organic field-effect transistors and complementary inverter based on low-temperature processed Al_2O_3 dielectric. Organic Electronics [Internet]. 2016; **34**:118-123. DOI: 10.1016/j.orgel.2016.04.023

[56] Zhang X-H, Domercq B, Wang X, Yoo S, Kondo T, Wang ZL, et al. High-performance pentacene field-effect transistors using Al_2O_3 gate dielectrics prepared by atomic layer deposition (ALD). Organic Electronics. 2007;**8**(6): 718-726

[57] Urasinska-Wojcik B, Cocherel N, Wilson R, Burroughes J, Opoku J, Turner ML, et al. 1Volt organic transistors with mixed self-assembled monolayer/Al_2O_3 gate dielectrics. Organic Electronics [Internet]. 2015;**26**: 20-24. DOI: 10.1016/j.orgel.2015.07.009

[58] Ezhilvalavan S, Tseng TY. Preparation and properties of tantalum pentoxide (Ta_2O_5) thin films for ultra large scale integrated circuits (ULSIs) application—A review. Journal of Materials Science: Materials in Electronics [Internet]. 1999;**10**(1):9-31. DOI: 10.1023/A:1008970922635

[59] Atanassova E. Thin RF sputtered and thermal Ta_2O_5 on Si for high density DRAM application. Microelectronics and Reliability. 1999;**39**(8):1185-1217

[60] Sakai H, Furukawa Y, Fujiwara E, Tada H. Low-voltage organic field-effect transistors with a gate insulator of

Ta_2O_5 formed by sputtering. Chemistry Letters. 2004;**33**(9):1172-1173

[61] Chiu CJ, Chang SP, Chang SJ. High-performance A-IGZO thin-film transistor using Ta_2O_5 gate dielectric. IEEE Electron Device Letters. 2010; **31**(11):1245-1247

[62] Bartic C, Jansen H, Campitelli A, Borghs S. Ta_2O_5 as gate dielectric material for low-voltage organic thin-film transistors. Organic Electronics: Physics, Materials, and Applications. 2002;**3**(2):65-72. DOI: 10.1016/S1566-1199(02)00034-4

[63] Kumaki D, Yahiro M, Inoue Y, Tokito S. Improvement of mobility and stability in pentacene-TFT by chemical surface treatment. Journal of Photopolymer Science and Technology. 2006;**19**(1):41-44

[64] Wang X, Xu J-B, Wang C, Du J, Xie W. High-performance graphene devices on SiO_2/Si substrate modified by highly ordered self-assembled monolayers. Advanced Materials [Internet]. 2011;**23**(21):2464-2468. DOI: 10.1002/adma.201100476

[65] Jinno H, Yokota T, Matsuhisa N, Kaltenbrunner M, Tachibana Y, Someya T. Low operating voltage organic transistors and circuits with anodic titanium oxide and phosphonic acid self-assembled monolayer dielectrics. Organic Electronics: Physics, Materials, and Applications [Internet]. 2017;**40**:58-64. DOI: 10.1016/j.orgel.2016.10.034

[66] Shin YC, Dresselhaus MS, Kong J. Chapter 4: Preparation of graphene with large area. In: Tanaka K, SBT I, editors. Carbon Nanotubes and Graphene. 2nd ed. Oxford: Elsevier; 2014. pp. 39-76

[67] Gala F, Zollo G. Dielectric properties of self-assembled monolayer coatings on a (111) silicon surface. Journal of Physical Chemistry C. 2015; **119**(13):7264-7274

[68] Liu X, Guo Y, Ma Y, Chen H, Mao Z, Wang H, et al. Flexible, low-voltage and high-performance polymer thin-film transistors and their application in photo/thermal detectors. Advanced Materials [Internet]. 2014; **26**(22):3631-3636. DOI: 10.1002/adma.201306084

[69] Mohammadian N, Faraji S, Sagar S, Das BC, Turner ML, Majewski LA. One-volt, solution-processed organic transistors with self-assembled monolayer-Ta_2O_5 gate dielectrics. Materials (Basel). 2019;**12**(16):2563

[70] He W, Xu W, Peng Q, Liu C, Zhou G, Wu S, et al. Surface modification on solution processable ZrO_2 high-k dielectrics for low voltage operations of organic thin film transistors. The Journal of Physical Chemistry C [Internet]. 2016;**120**(18): 9949-9957. DOI: 10.1021/acs.jpcc.6b03638

[71] Kim CS, Jo SJ, Lee SW, Kim WJ, Baik HK, Lee SJ. Surface-modified high-k oxide gate dielectrics for low-voltage high-performance pentacene thin-film transistors. Advanced Functional Materials [Internet]. 2007;**17**(6): 958-962. DOI: 10.1002/adfm.200600747

[72] Hsiao R, Miller D, Kellock A. Response surface study of resist etching in high density oxygen plasma and interactions of O_2 plasma with NiFe, Cu, Ta, and Al_2O_3. Journal of Vacuum Science & Technology A [Internet]. 1996;**14**(3):1028-1032. DOI: 10.1116/1.580127

[73] Niimi H, Koh K, Lucovsky G. Ultra-thin gate dielectrics prepared by low-temperature remote plasma-assisted oxidation. Nuclear Instruments and Methods in Physics Research, Section B: Beam Interactions with Materials and Atoms [Internet]. 1997;**127–128**:

364-368. DOI: 10.1016/S0168-583X(96)00958-5

[74] Zhang F, Di C, Berdunov N, Hu Y, Hu Y, Gao X, et al. Ultrathin film organic transistors: Precise control of semiconductor thickness via spin-coating. Advanced Materials. 2013; **25**(10):1401-1407

[75] Sloppy JD. Anodization Mechanism and Properties of Bi-Layer Tantalum Oxide Formed in Phosphoric Acid. Pennsylvania, US: The Pennsylvania State University; 2009

[76] Deman AL, Erouel M, Lallemand D, Phaner-Goutorbe M, Lang P, Tardy J. Growth related properties of pentacene thin film transistors with different gate dielectrics. Journal of Non-Crystalline Solids. 2008;**354**(15–16):1598-1607. DOI: 10.1016/j.jnoncrysol.2007.10.027

[77] Zhou J, Yang K, Zhou J, Liu Y, Peng J, Cao Y. Poly (3-hexylthiophene) thin-film transistors with dual insulator layers. Japanese Journal of Applied Physics. 2007;**46**(3R):913. DOI: 10.1143/JJAP.46.913

[78] Majewski LA, Schroeder R, Grell M. One volt organic transistor. Advanced Materials. 2005;**17**(2):192-196

[79] Feng L, Cui Q, Zhao J, Tang W, Guo X, Zou J, et al. Preparation of tantalum anodic oxide film in citric acid solution—Evidence and effects of citrate anion incorporation. IEEE Electron Device Letters [Internet]. 2017; **17**(1):178-183. DOI: 10.5229/jecst.2013.4.4.163

[80] Mohammadian N, Das BC, Majewski LA. Low-Voltage IGZO TFTs using solution-deposited OTS-modified Ta_2O_5 dielectric. IEEE Transactions on Electron Devices. DOI: 10.1109/TED.2020.2976634

Chapter 6

Tackling the Problem of Dangerous Radiation Levels with Organic Field-Effect Transistors

Irina Valitova, Zhihui Yi and Jonathan Sayago

Abstract

Accurate, quantitative measurements of ionizing radiation, commonly employed in medical diagnostic and therapeutic applications are essential prerequisites to minimize exposure risks. Common examples of radiation detectors include ionization chambers, thermoluminescent dosimeters, and various semiconductor detectors. Semiconductor dosimeters such as p/n type silicon diodes and MOSFETs have found widespread adoption due to their high sensitivity and easy processing. A significant limitation of these devices, however, is their lack of tissue equivalence. The high atomic number (relative to soft tissue) of silicon causes these devices to over-respond to photon beams that include a significant low energy component, for example, 1–10 kV, due to an enhanced photoelectric interaction coefficient. Organic field effect transistors (OFETs) are capable of providing tissue equivalent response to ionizing radiation in order to monitor more accurately the risk of exposure in medical treatments. This chapter presents the possibility to use different types of OFETs as ionizing and X-ray radiation dosimeters in medical applications.

Keywords: OFET dosimeters, ionizing radiation, X-ray detection, ion sensors, ionizing radiation dosimeters, radiation dosimeters

1. Introduction

Since the beginning of research in both radiotherapy and X-ray imaging, an attempt has been made to measure the cumulative exposure to radiation that could be harmful to the patient. Ionizing radiation is related to heavy charged alpha- and beta- particles, also X-rays which are highly energetic electromagnetic waves (light), can interact with matter to produce secondary ionizing radiation which can result in health issues. This is particularly the case in medical screening procedures such as mammography, tomographic, and nuclear imaging [1].

The need for high precision radiation monitoring is particularly important in radiotherapy. Approximately half of all patients diagnosed with cancer receive some form of radiotherapy over the course of their treatment [2]. In radiotherapy, a precise dose of radiation is delivered to control or eliminate the disease in neoplastic tissue but not large enough to incur in significant risk to damage healthy tissue. In such cases the risk of the procedure has to be carefully balanced against the expected benefits. Therefore, accurate dosimetry is needed to deliver successful radiotherapy.

Ionizing radiation sensors are devices that respond to ionic radiation and output a proportionally strong electrical or optical signal. To understand the performance of a sensor it is important to have an understanding of how the signals are generated and recorded.

The main quantities of dosimetry will be introduced in this chapter as well as the basic OFET sensing mechanism and a review of the most prominent research up to date in the field.

1.1 Sources of ionizing radiation

The source of ionizing radiation can originate from radiation generators (accelerators of charged particles and nuclear reactors), radioactive materials and electromagnetic radiation. The maximum attainable energy and intensity of ionizing radiation beams are related to the source of radiation. For example, cobalt-60 isotope emits gamma rays with energies of the gamma quantum at 1.17 and 1.33 MeV [3]. The radiation from radioactive materials has an energy spectrum that changes as a function of time due to the short half-life of some radioactive isotopes.

Sources of electron- or beta- rays can be different types of electron accelerators with energies ranging from several tens of MeV. Commonly used accelerators are the Van der Graff generators, linear accelerators and cyclotrons.

Electromagnetic radiation, X- and gamma rays, have the highest energy photons (light) in the electromagnetic spectrum and concern biomedical sensors because they can attain energies of 1 kV to 25 MeV, large enough to interact with its medium to form secondary charged particles that can be detected in a dosimeter.

1.2 Exposure, dose and dose equivalent

An ionizing radiation beam can be characterized by its exposure, which is the ratio of charge it creates in its medium to the air mass (SI: $C\ Kg^{-1}$). For convenience, the medium is air because air has an atomic number of 7.6, which is tissue equivalent (tissue has an atomic number of 7.4). The SI unit is the coulomb per kilogram, however, historically a common unit of measure is the roentgen (R). One R equals 0.000258 $C\ Kg^{-1}$.

The absorption characteristic of the target material (i.e. skin tissue) results in the concept of absorbed dose. The absorbed dose is the amount of energy absorbed per unit mass of target material (SI: $J\ Kg^{-1}$), as a result of an exposure to ionizing radiation. Common units of measure for absorbed dose are the rad, the gray (Gy) and the erg; 1 rad = 0.01 Gy = 0.01 $J\ Kg^{-1}$ = 100 ergs g^{-1}.

In other terms, exposure refers to the radiation in an area and the dose is the amount of that radiation is expected to be absorbed by a person or a target material. For a given radiation source, the absorbed dose will depend on the matter that absorbs the radiation. For example, for an exposure of 1 R of gamma rays with an energy of 1 MeV, the dose in silicon will be 0.88 rad and the dose in human tissue will be 1 rad [4].

The absorbed dose itself is ineffective for describing the biological effects of radiation, in which, to a close approximation an equal amount of absorbed energy corresponds to an equal amount of damage [4]. For this purpose, the dose equivalent was defined, as the product of the absorbed dose times the radiation weighting factor taking into account the beam type. **Table 1** summarizes the weighting factor (w) for different types of radiation [5].

Although w is dimensionless, the unit employed to define dose equivalent is not the same than that for dose (Gy) but is the sievert (Sv): 1Sv = 1 $J\ Kg^{-1}$. People at risk for repeated ionizing radiation exposure are commonly monitored and restricted to effective doses of 100 mSv every 5 years with a maximum of 50 mSv allowed in any given year [6].

Type of radiation	Weighting factor (w)
Photons	1
Electrons	1
Neutrons	
<10 keV	10
10–100 keV	20
>100 keV to 2 MeV	10
>2–20 MeV	5
>20 MeV	5
Protons, energy >2 MeV	5
α-particles	20

Table 1.
Radiation weighting factor (w) for different types of radiation beams [5].

1.3 Interaction of ionizing radiation with matter

Ionizing radiation can be classified into heavy- protons, ions- and light charged particles- electrons and positrons. The type of radiation clearly modifies its interaction with matter. As one would imagine, a heavy charged particle has a larger mean range or penetration depth than lighter particles. Both particles can loose energy through collisional or radiative energy transfer [5]. The first mechanism dominates in heavy particles and the latter is mainly present in electrons passing through high-Z materials.

Most biomedical sensing applications involve collision dissipation of low-energy ion beams (i.e., 10 keV) thus the penetration depth upon collision is relatively low yet the thickness or volume of the sensing material should be large enough to ensure that most energy associated with the ionizing particle is deposited within the active layer. Of course, the penetration depth also depends on the material the beam is interacting with.

The four most common interaction types are Rayleigh scatter, photoelectric effect, Compton scatter, and electron-positron pair production [7]. The main photon- medium interactions are in the energy range from 100 kV to 18 MV (produced by medical accelerators for cancer treatment) are the photoelectric effect and Compton scattering. Rayleigh scattering and pair production are responsible for a lower percentage of photon interactions.

Rayleigh scattering is a process where a photon interacts with an orbital electron in the medium and deflects through a very small angle. The angle of deflection is proportional to the energy loss. So the Rayleigh scattering is an elastic process, meaning that very small amount or no energy is lost by the scattered photon.

The mass attenuation or extinction coefficient for Rayleigh scattering, σ_R/ρ, depends on the energy of the photon and the atomic number of the material through which the photon is traveling:

$$\sigma_R/\rho \propto Z/E^2, \quad (1)$$

where E is the energy of the incident photon and Z is the atomic number of the medium. Pair production occurs when the photon passes close to the nucleus of the atoms and if the energy of photon is high enough (>1.022 MeV) an electron and positron pair will be created. The mass attenuation coefficient for pair production, κ/ρ, depends on the atomic number of the material through which the photon is traveling: $\kappa/\rho \propto Z$.

The Compton collisions become predominant when the photon energy becomes significantly larger than the binding energy of electron (i.e., 50 keV and above). The photon transfers part of its energy to the electron and scatters.

The photoelectric effect occurs when a photon's energy exceeds the binding energy of the electron on its shell. In this case, the electron gets ejected from the atom with a kinetic energy equivalent to the incident photon energy minus the binding energy of the electron. This photoelectric current (the number of photoelectrons) strongly depends on the atomic number Z of the target, while Compton effect does not. This is why the photoelectric current from high Z materials is higher than that of tissue equivalent materials with lower Z (7.4).

Commonly employed sensing materials are high-Z (high atomic number) which include carbon, silicon, germanium, cesium iodide, sodium iodide. However, they have limitations in the processing of large area pixelated sensor matrices due to the relatively low resolutions achievable with these devices and their low life cycles due to structural defects caused in the sensing material by the ionizing radiation beam.

2. Detectors of ionizing radiation

A dosimeter is a device that measures dose uptake of ionizing radiation beams. The response of an ideal dosimeter should be: linear with the absorbed dose and consistent over time; directly proportional to the dose and independent of the type of radiation and its energy; independent of the pulsed and continuous dose rates and independent of the average dose rate; proportional to the absorbed dose and independent of the incident radiation angle; reproducible and stable [5].

There are a few standard methods for determining the exposure or absorbed dose [4]. The calorimetric method, or thermodynamic method, measures the total heat generated during the absorption of energy from a radiation field. Alternatively, an ionization chamber can be used to measure the number of ions produced in air (or other gases) by charged particles or electromagnetic radiation. The device is fairly simple; it essentially consists of two electrodes in a gas-filled space in which the incident radiation produces ionization. A voltage applied between the two electrodes draws the ions to them and the resulting current flow is measured.

When measuring high energy radiation (MeV), chemical changes in matter (polymerization, oxidation, reduction, degradation) as well as physical changes in material properties (color) as a result of radiation, can be used to quantify radiation. The best example is the viscometric dosimeter that measures the change in molecular structure of polymethylmethacrylate polymer as it degrades upon radiation. The degradation results in a decrease of its average molecular weight which then can be measured by dissolving it and determining the solution viscosity. Another good example is the glass dosimeter in which high doses of radiation produce measurable darkening of the glass by the formation of absorption bands in the electronic structure of the material. Changes in the optical properties are determined with a spectrophotometer and compared with an unexposed reference glass [8]. Polyethylene releases hydrogen under irradiation, which in turn can increase the relative pressure of a sealed chamber providing a cost-effective and versatile dosimeter solution (in terms of the sensor location, close or far away from the source) for high dose sensing.

One difficulty of employing organic materials as sensing material is the requirement of a thick layer of material to absorb a significant portion of the radiation because the exposure is a function of the source location. The use of the viscometric dosimeter has the advantages of using organic tissue-equivalent material as sensing material and being cost-effective simple solution, however the accuracy is only about 20%.

2.1 Solid-state methods for sensing ionizing radiation

Solid-state dosimeters are well established today. They commonly employ inorganic semiconductors with a small number of impurities as the sensing material.

On one hand, ionizing radiation induces electronic states (traps) at intermediate energies between the valence and conduction bands of the semiconductor. If the trap energy levels are large enough to produce light when excited electrons decay, the characteristic light emitted through this process is proportional to the number of traps and thus, proportional to the dose absorbed by the material upon ionizing radiation. An external source of light can be applied to excite electrons from the valence band to the conduction band, see **Figure 1**, and a spectrophotometer can be used to detect the light emitted from the photon as it decays through the trap.

2.2 MOSFET sensors

Metal oxide semiconductor field-effect transistors are well known and have been used for decades as dosimeters. They consist of a semiconducting channel sandwiched between drain and source electrodes (electricity can be thought of as water flowing through a pipeline) and switched on/off by a gate consisting of a dielectric and a third electrode, **Figure 2**.

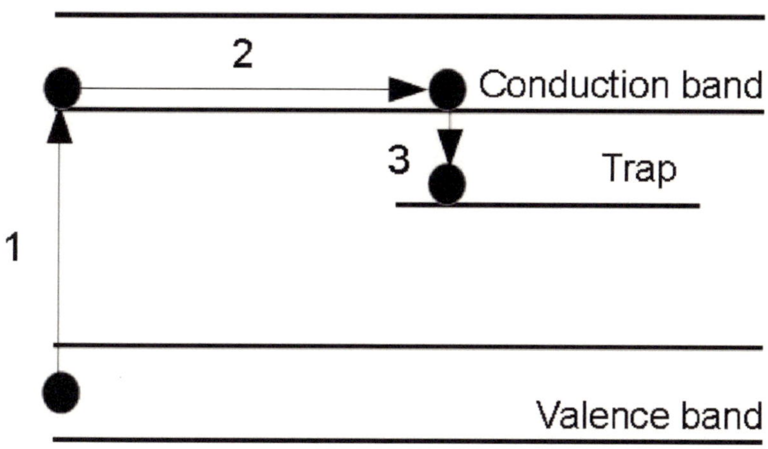

Figure 1.
Optically stimulated luminescence for dosimetry: An electron (1) is excited by a light source into the conduction band, (2) migrates through it and (3) falls into a trap emitting a characteristic light proportional to the number of traps previously induced by ionizing radiation.

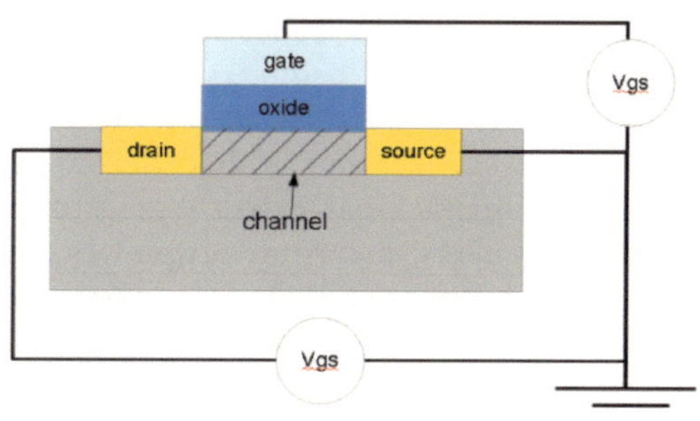

Figure 2.
A metal oxide semiconductor field-effect transistor typical structure.

The flow of current between drain and source electrodes is limited by the number of charge carriers available in the semiconductor channel. Applying a gate voltage polarizes the dielectric material and results in the accumulation of charge carriers at the interface between the semiconductor and the dielectric thus increasing the conductivity in that region (field-effect). Transistors are typically characterized by the threshold voltage one must apply to the gate to switch on a drain-source current. When the device is exposed to ionizing radiation, a fraction of the charge carriers are trapped in the oxide-substrate interface creating an electric field that increases the value of the threshold voltage. The variation in threshold voltage is therefore directly proportional to the traps created in the semiconducting channel by the delivered dose.

A slight variation of the MOSFET dosimeters can be found in p-n or p-i-n diodes in which the degradation of the semiconductor by ionizing radiation results in differences in time-resolved photoconductivity.

Commonly utilized materials in solid-state dosimeters are LiF in combination with Mg, Cu, P and Al_2O_3, for low-dose dosimetry [9, 10]. Sodium iodide doped with thallium (NaI:Tl) is probably the most used dosimeter for gamma detection [11].

The advantages of solid-state dosimeters include high sensitivity, high reproducibility and linearity over a wide range. However, their main disadvantage is related to the lack of tissue equivalence.

3. Organic electronics and their relevance for tissue equivalences

In order to track correctly the mechanisms of energy release in skin tissue, the ideal dosimeter must be tissue equivalent, with a Z as close as possible to 7.4. At high energy sensing, tissue equivalence may be even more relevant due to the changes in the effective cross section particle interaction with the target originated from resonance peaks in elastic scattering [5]. In tissue equivalent sensors, the mass stopping power of the target varies slowly as a function of the electron energy, therefore, there are no sharp peaks commonly observed with resonances with high-energy particles and high Z materials.

Organic electronics, based on *soft* materials, have gained interest in radiation dosimetry for tissue equivalent applications. The low atomic number of organic materials is comparable to that of average human soft tissues, which suggests that these devices may require fewer or smaller correction factors to translate a measured signal into a dose absorbed by skin tissue under conditions that deviate significantly from calibration conditions. These devices offer other advantages over their inorganic counterparts, namely—their versatility, mechanical flexibility, solution processability, low cost and suitability for large and nanoscale applications.

Changes of both electronic and optical properties of organic materials under irradiation can be used for high accuracy dosimetry. The shift of organic polymers absorbance or reflectance allows visualizing structural defects due to ionizing radiation [12], and electronic changes can be detected in different device configurations, such as capacitors, diodes and Organic Field-Effect Transistors (OFETs) [13–15].

A radiation-induced changes in the electronic properties of insulating polymer materials like poly(methyl methacrylate) (PMMA), polystyrene (PS) or polyethylene has been known for decades [16] in which the radiation-produced free electrons trapped in the band gap are thermally released into the conduction band under the radiation field. Later the use of organic semiconducting polymers helped to improve the dosimeter sensitivity due to the better charge carrier mobility and better carrier lifetime. The use of organic polymers in different organic electronic devices such as organic diodes and Organic Field-Effect Transistors (OFETs) with the tissue equivalence that we previously discussed will be introduced further.

3.1 Dosimetry based on organic field-effect transistors

Various parameters, such as off current, on current, threshold voltage V_{th}, current ratio and subthreshold swing, can be extracted as a function of ionizing radiation dose from the measured transfer characteristics of organic field effect transistors. Raval et al. introduced bottom gate top contact OFET with P3HT as semiconductor on a silicon dioxide (SiO_2) dielectric layer as radiation dosimeter. After ϒ-irradiation up to 41 krad using a Co-60 radiation source, they found a decrease in the on current by a factor of 2, an increase in off current by a factor of 150, and a decrease in the mobility. The threshold voltage was shifted negatively due to positive charge accumulation in the silicon dioxide [17]. Later the same group investigated Pentacene OFET using similar device structure shown in **Figure 3**. The silicon nitride as a passivation layer was used to protect organic materials from

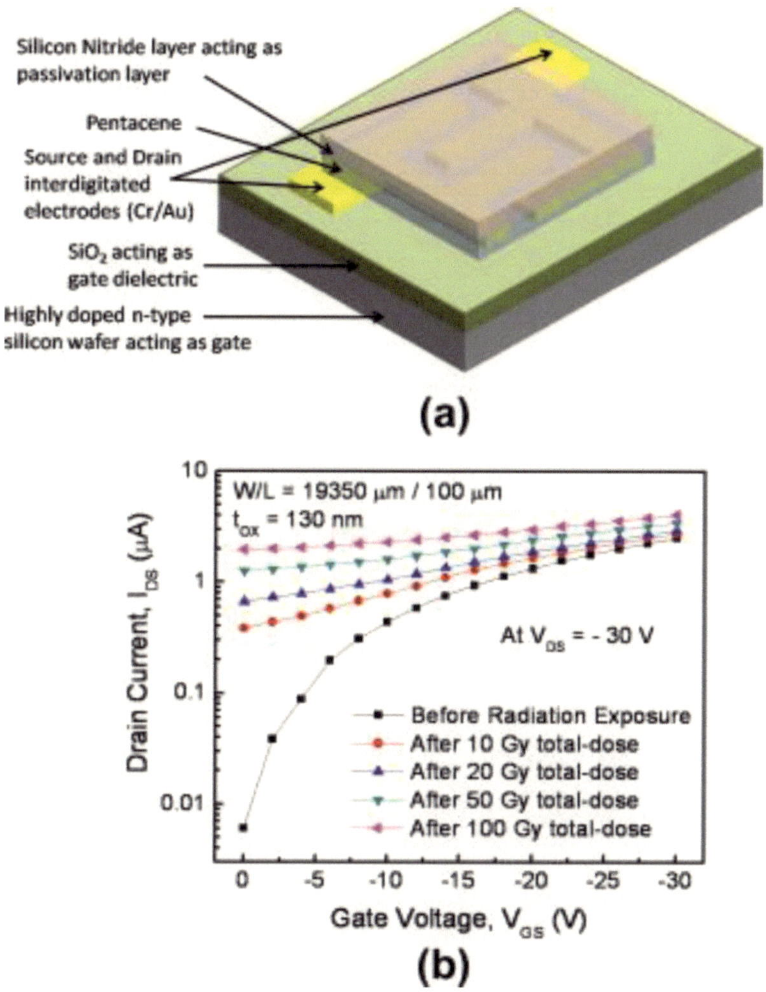

Figure 3.
(a) Schematic of a pentacene OFET in bottom gate bottom contact configuration with interdigitated source and drain electrodes using a silicon nitride passivation layer, and (b) I_d-V_g characteristics for a Pentacene OFET with exposure to increasing dose of ionizing radiation [18].

interaction with air. After exposure to a total of 100 Gy dose of ionizing radiation off current was increased 320 times which resulted in a sensitivity of 20 nA/Gy. The threshold voltage shift resulted in a sensitivity of 0.3 V/Gy [18].

Furthermore, they introduced CuPc OFET dosimeter after γ-irradiation with a minimum dose of 10 rad going up to a maximum dose of 100 krad. To solve the resolution limitation of measuring the off current for less than 1 krad total-dose exposure three and five OFETs were connected in parallel. The transfer characteristics of those sensors after different total-dose exposures to γ-radiation are shown in **Figure 4**. The measured sensitivity from off current shifts after irradiation was

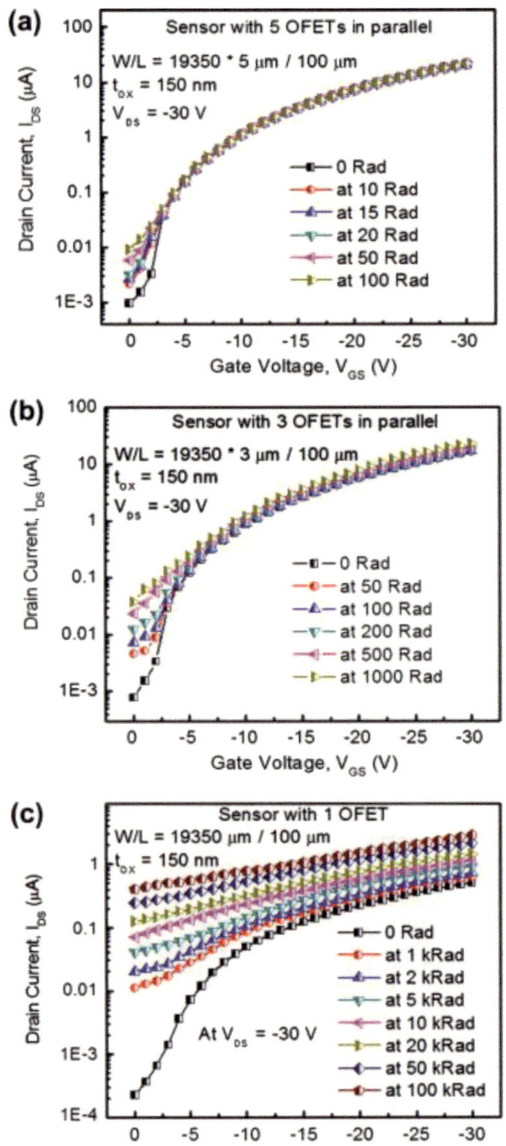

Figure 4.
(a) Transfer characteristics for a sensor with five CuPc-OFETs each with W/L = 19,350 lm/100 lm stacked in parallel for total-dose exposures up to 100 rad with the minimum dose of 10 rad. (b) Transfer characteristics for a sensor with three CuPc OFETs each with W/L = 19,350 μm/100 μm stacked in parallel for total-dose exposures up to 1 krad with the minimum dose exposure of 50 rad. (c) Transfer characteristics for a CuPc OFET sensor with W/L = 19,350 μm/100 μm exposed to a minimum of 1 krad, and with increasing total-dose exposures up to 100 krad [19].

of 0.02 A/rad. From the V_{th} shift measured at constant drain current $1e^{-7}$ A the sensitivity of 1.5 10^4/rad was observed for a total of 100 krad dose-exposure [19].

Kim et al. introduced a rubrene semiconductor OFET as a dosimeter to electron beam irradiation [15]. To show that radiation induced charges can be trapped not only in SiO_2 dielectric and Si/SiO_2 interface but also in organic semiconductor, they compared two sets of devices. In one set of devices they irradiated a silicon/silicon dioxide substrate before deposition of a rubrene semiconductor. The on and off currents were about the same while the mobility fell by about 50% after 10^7 rad in comparison to pre-irradiation conditions. A second set of devices was irradiated after the deposition of rubrene. The mobility decrease of more than 50% was found only after 10^5 rad dose exposure. Moreover, the subthreshold swing was decreased with increased radiation dose. So the charge trap density at rubrene/SiO_2 interface was increased as a function of radiation dose. They concluded that electrons could induce traps not only on interface but also in the bulk semiconductor.

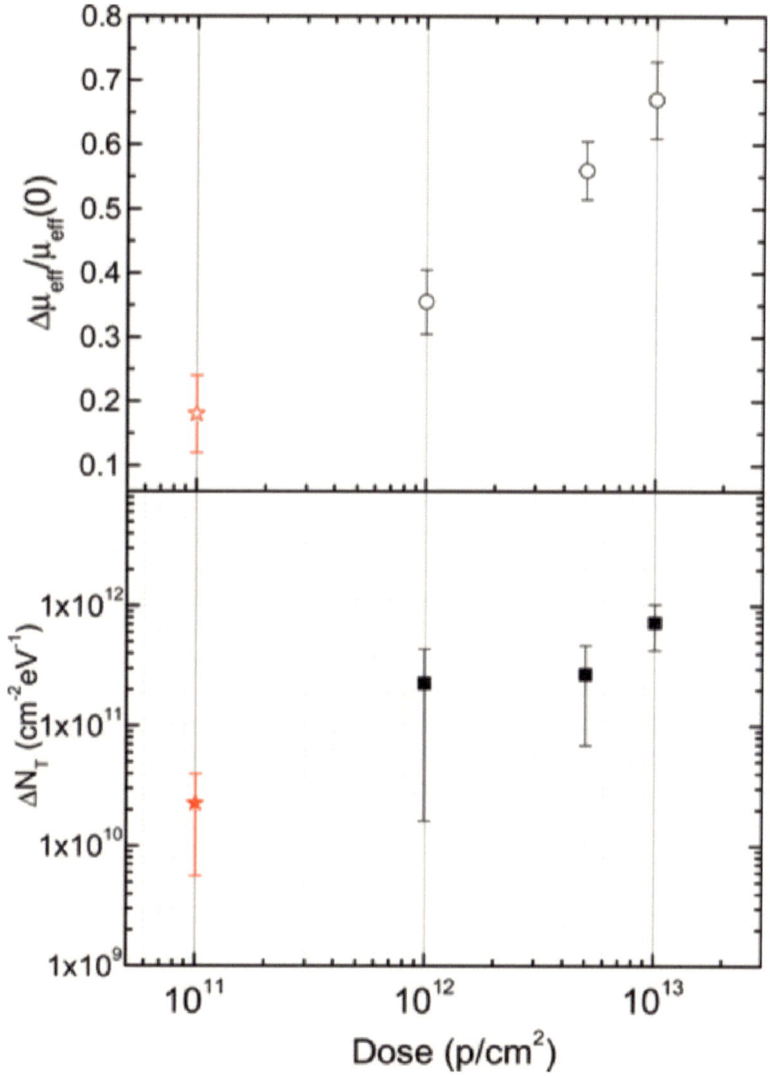

Figure 5.
Variation of the trap densities, ΔN_T (bottom solid squares), and of the channel mobility relative to $\mu_{eff}(0)$ and $\Delta\ \mu_{eff}/\mu_{eff}$ (0) (top open circles) as a function of the irradiation dose [20].

The reduction of charge carrier mobility proves that radiation induced density of traps also energetically located near the highest occupied molecular orbital of the organic semiconductor. However, Basirico and colleagues only partially related the reduction of charge carrier mobility with increase of charge carrier traps in 6,13-bis(triisopropylsilylethynyl)-pentacene (TIPS-pentacene)-based field effect transistors with organic dielectric irradiated by high energy protons [20]. The high energy protons induce defects in the organic dielectric and strains in the TIPS-pentacene layer leading to mobility reduction. The variation of trap densities and charge carrier mobilities as a function of radiation dose shown in **Figure 5**.

Jain et al. improved the response of OFET to radiation by mixing organic semiconductor with Polystyrene probably increasing the amount of trap carrier density. TIPS-Pentacene (TP) and Polystyrene (PS) blend as semiconductor material in OFET for sensing gamma rays from cobalt-60 (^{60}Co) radiation source [21]. The device configuration was n-type Si as a gate electrode, 100 nm of SiO_2 as a gate dielectric layer, TP/PS blend as semiconductor and interdigitated Au source/drain electrodes on the top. Devices irradiated before deposition of TIPS-Pentacene and devices irradiated after deposition of semiconductor were compared to show the amount of charges trapped in TP/PS blend and in the SiO_2/TP-PS interface. It was shown that interface trap density of SiO_2/TIPS-Pentacene was significantly higher in irradiated TIPS-Pentacene than trap density of devices irradiated before TIPS-Pentacene deposition. The sensitivity of TIPS-Pentacene transistors was highest among similar organic transistors in the literature, 3 V/Gy [21].

3.2 Dosimetry based on organic diodes and single layered structures

Boroumand et al. reported the tissue-equivalent direct response of organic semiconducting polymers, such as poly(2-methoxy-5-(2′-ethylhexyloxy)-p-phenylene vinylene) (MEH-PPV) or poly 9,9-dioctylfluorene (PFO), to 17 keV X-rays from a 50 kVp molybdenum source. In order to maximize the X-ray photon attenuation, the typical polymer film thickness was approximately 20 μm. The sensitivities of the devices at −10 V applied bias were 0.064 nC/mGy for PFO and 0.1 nC/mGy for MEH-PPV, and 0.24 nC/mGy at −50 V for PFO. These values correspond to sensitivities per unit volume of 128–480 nC/mGy/cm^3, which are similar to silicon-based devices [22].

Flexible dosimeters incorporating a ~ 10 μm poly([9,9-dioctylfluorenyl-2,7-diyl]-co-bithiophene) (F8T2) film were fabricated on a polyimide substrate [23]. The higher electric field in these devices helped to separate X-ray-generated charge carriers leading to increase the sensitivity from 54.2 to 158.2 nC/mGy/cm^3 when reversed applied voltages were −10 and −50 V, respectively.

Intaniwet demonstrated the influence of charge carrier mobility in organic semiconductors on the dosimeter sensitivity [24]. The blend of high mobility 6,13-Bis (triisopropylsilylethynyl)pentacene (TIPS-pentacene) with conjugated polymer poly(triarylamine) (PTAA) was used to improve the sensitivity of organic dosimeters to detect up to 17.5 keV X-rays from a molybdenum source. The PTAA device with charge carrier mobility of $1.3*10^{-6}$ cm^{2}V^{-1} s^{-1} possessed a sensitivity of 116 nC/mGy/cm^3 and the TIPS:PTAA blend (17:1) with mobility $19*10^{-6}$ cm^2/Vs had four times more sensitivity, up to 457 nC/mGy/cm^3. The sensitivity of organic semiconducting single crystal (OSSC) dosimeters did not appear to depend on the charge carrier mobility. For example, it was shown that the high mobility single crystal ruberene possessed poorer X-ray sensitivity than the low mobility DNN (1,5-dinitronaphthalene) [25].

Another important factor known to affect organic based dosimeters is the morphology of the semiconducting film. Dr. Fraboni studied the influence of anisotropic π-π stacking of molecules and the photo-response when exposed to X-ray radiation [26]. They selected 1,5-dinitronaphthalene (DNN) as the active material which has strong π-π stacking in one axis of the crystal and compared the results with 4-hydroxycyanobenzene (4HCB) with a more planar morphology. The photoconductivity increased almost linearly with the applied voltage bias to the crystal along the vertical axis while it tends to saturate along the planar ones even for driving voltages as low as 50 V with no hysteresis effect and considerable reproducibility and device life-cycles, indicating that OSSCs are promising candidates for direct X-ray radiation detection (**Figure 6**, left). They reported that the

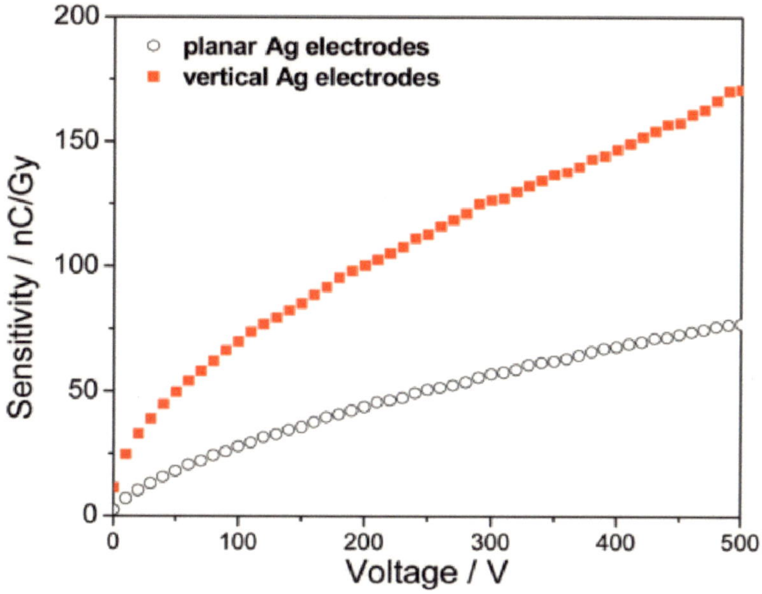

Figure 6.
Comparison between the sensitivity values at different bias voltages for a 4HCB- based detector under 35 keV X-ray irradiation in the planar (black circles) and vertical (red squares) electrode configuration [26].

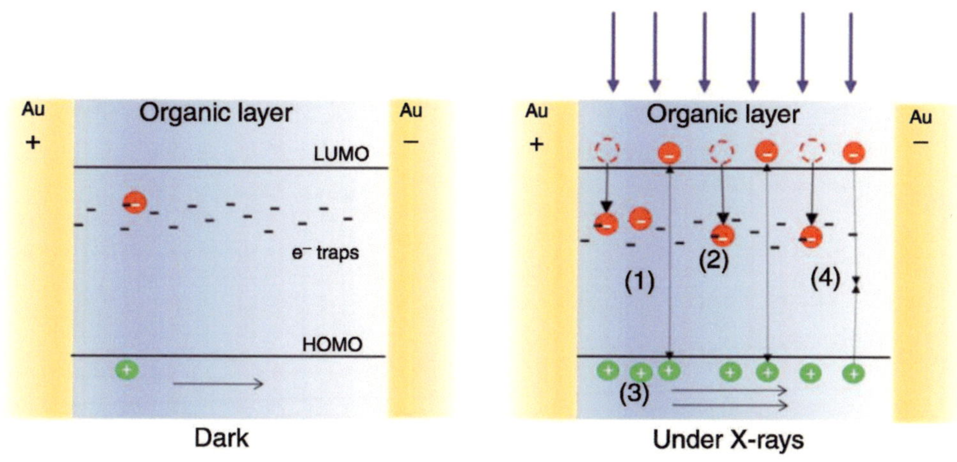

Figure 7.
Schematic of the process of modulation of the conductivity induced by X-rays exposure of TIPS-pentacene thin films: (left) in dark, the conductivity is mainly due to the intrinsic charge carriers; (right) under X-ray irradiation: (1) additional electrons and holes are created and holes drift along the electric field until they reach the collecting electrode while (2) electrons may remain trapped in deep trap states within the organic material. (3) to guarantee charge neutrality, holes are continuously emitted from the injecting electrode. For each electron–hole pair created upon radiation, more than one hole contributes to the photocurrent leading to a photoconductive gain effect. (4) recombination process takes place counterbalancing the charge photogeneration in the steady-state [28].

film thickness of crystals has minor influence on the sensitivity for a device with electrodes larger than 2 mm^2. The maximum obtained sensitivity for large electrode area samples is 175 nC Gy^{-1} [27].

Fraboni and co-workers proposed a model to explain the photocurrent signal induced by X-ray radiation (**Figure** 7) [28]. The model described the accumulation of holes in the LUMO level with the X-ray radiation. Radicals generated from the X-rays radiation activated energy levels within the HOMO and LUMO levels where electrons were trapped. The induced holes in the LUMO level induce an increase in the photoconductivity easily measurable due to the high sensitivity to X-rays and electrical response of the material.

4. Conclusions

Organic field effect transistors are new but very potential devices to be used as ionization radiation dosimeters. Different parameters like off current, on current, threshold voltage, current ratio and sub-threshold swing can be monitored as a function of ionizing radiation dose. Organic field effect transistors are capable of providing a tissue equivalent response to ionizing radiation. High Z counterparts, such as insulating layer and metal electrodes can be fabricated making use of organic dielectrics and conducting polymers (PEDOT:PSS or others), respectively. Moreover, organic materials possess outstanding properties, such as large area processing on lightweight and flexible substrates and ability to chemically tailor their properties. In particular, lightweight, flexible OFETs can be attached to the patient during the radiation treatment with possibility to read out parameters after treatment or operating transistors at very low voltages. For sure more work should be done on sensitivity improvement and characterization of such dosimeters. For example, it is important to check their linearity, energy and dose rate dependence, and so on. The sensitivity of such dosimeters can be improved by increasing the trap carrier density in organic semiconductor or dielectric layers. Similar to organic memory devices dielectric layer can be complex or with floating gate to store more charges and thus show the improved response to ionizing radiation.

Author details

Irina Valitova[1], Zhihui Yi[2] and Jonathan Sayago[3]*

1 Department of Physics and Atmospheric Science, Dalhousie University, Halifax, Nova Scotia, Canada

2 PhDs Network Inc., Toronto, Canada

3 Instituto de Energías Renovables, Universidad Nacional Autónoma de México, Temixco, Morelos, México

*Address all correspondence to: jjsayago@ier.unam.mx

References

[1] Fazel R, Krumholz HM, Wang Y, Ross JS, Chen J, Ting HH, et al. Exposure to low-dose ionizing radiation from medical imaging procedures. The New England Journal of Medicine. 2009;**361**:849-857

[2] Symonds PR, Deehan C, Meredith C, Mills JA. Walter and Miller's Textbook of Radiotherapy E-book: Radiation Physics, Therapy and Oncology. UK: Elsevier Health Sciences; 2012

[3] Tsividis Y, McAndrew C. Operation and Modeling of the MOS Transistor. Available from: http://cds.cern.ch/record/1546736 [Accessed: 09 April 2015]

[4] Van de Voorde MH. Effects of Radiation on Materials and Components: Megarad Dosimetry. Geneva (Switzerland): European Organization for Nuclear Research; 1969

[5] Jones DP. Biomedical Sensors. NY: Momentum Press; 2010

[6] Wrixon AD. New ICRP recommendations. Journal of Radiological Protection. 2008;**28**:161-168

[7] Attix FH. Introduction to Radiological Physics and Radiation Dosimetry. Weinheim: John Wiley & Sons; 2008

[8] Ezz-Eldin FM, Abdel-Rehim F, Abdel-Azim AA, Ahmed AA. Soda-lime-silica glass for radiation dosimetry. Medical Physics. 1994;**21**:1085-1089

[9] mez-Ros JMG, ez-Vergara JCS, Romero AM, Budzanowski M. Fast automatic glow curve deconvolution of LiF: Mg, Cu, P curves and its application in routine dosimetry. Radiation Protection Dosimetry. 1999;**85**:249-252

[10] Duggan L, Sathiakumar C, Warren-Forward H, Symonds M, McConnell P, Smith T, et al. Suitability of LiF: Mg, Cu, P and Al_2O_3: C for low dose measurements in medical imaging. Radiation Protection Dosimetry. 1999;**85**:425-428

[11] Hofstadter R, Stein MH. Twenty five years of scintillation counting. IEEE Transactions on Nuclear Science. 1975;**22**:13-25

[12] Isa NM, Baharin R, Majid RA, Rahman WAWA. Optical properties of conjugated polymer: Review of its change mechanism for ionizing radiation sensor. Polymers for Advanced Technologies. 2017;**28**:1559-1571

[13] Korostynska O, Arshak K, Morris D, Arshak A, Jafer E. Radiation-induced changes in the electrical properties of carbon filled PVDF thick films. Materials Science and Engineering B. 2007;**141**:115-120

[14] Arshak A, Arshak K, Korostynska O, Zleetni S. Review of various gamma radiation dosimeters based on thin and thick films of metal oxides and polymer materials. 2003 IEEE Nuclear Science Symposium. Conference Record (IEEE Cat. No.03CH37515) Vol. 1. Portland, OR, USA: IEEE; 2003. pp. 78-82

[15] Kim JJ, Ha JM, Lee HM, Raza HS, Park JW, Cho SO. Effect of electron-beam irradiation on organic semiconductor and its application for transistor-based dosimeters. ACS Applied Materials & Interfaces. 2016;**8**:19192-19196

[16] Warner AJ, Muller FA, Nordlin HG. Electrical conductivity induced by ionizing radiation in some polymeric materials. Journal of Applied Physics. 1954;**25**:131-131

[17] Raval HN, Tiwari SP, Navan RR, Rao VR. Determining ionizing radiation

using sensors based on organic semiconducting material. Applied Physics Letters. 2009;**94**:123304

[18] Raval HN, Sutar DS, Nair PR, Ramgopal Rao V. Investigation of effects of ionizing radiation exposure on material properties of organic semiconducting oligomer – Pentacene. Organic Electronics. 2013;**14**:1467-1476

[19] Raval HN, Sutar DS, Ramgopal Rao V. Copper(II) phthalocyanine based organic electronic devices for ionizing radiation dosimetry applications. Organic Electronics. 2013;**14**:1281-1290

[20] Basiricò L, Basile AF, Cosseddu P, Gerardin S, Cramer T, Bagatin M, et al. Space environment effects on flexible, low-voltage organic thin-film transistors. ACS Applied Materials & Interfaces. 2017;**9**:35150-35158

[21] Jain S, Surya SG, Suggisetti PK, Gupta A, Rao VR. Sensitivity improvement of medical dosimeters using solution processed TIPS-Pentacene FETs. IEEE Sensors Journal. 2019;**19**:4428-4434

[22] Boroumand FA, Zhu M, Dalton AB, Keddie JL, Sellin PJ, Gutierrez JJ. Direct X-ray detection with conjugated polymer devices. Applied Physics Letters. 2007;**91**:033509

[23] Mills CA, Intaniwet A, Shkunov M, Keddie JL, Sellin PJ. 7449. 2009

[24] Intaniwet A, Mills CA, Shkunov M, Sellin PJ, Keddie JL. Heavy metallic oxide nanoparticles for enhanced sensitivity in semiconducting polymer X-ray detectors. Nanotechnology. 2012;**23**:235502

[25] Basiricò L, Ciavatti A, Sibilia M, Fraleoni-Morgera A, Trabattoni S, Sassella A, et al. Solid state organic X-ray detectors based on rubrene single crystals. IEEE Transactions on Nuclear Science. 2015;**62**:1791-1797

[26] Fraboni B, Ciavatti A, Basiricò L, Fraleoni-Morgera A. Organic semiconducting single crystals as solid-state sensors for ionizing radiation. Faraday Discussions. 2014;**174**:219-234

[27] Ciavatti A, Capria E, Fraleoni-Morgera A, Tromba G, Dreossi D, Sellin PJ, et al. Toward low-voltage and bendable X-ray direct detectors based on organic semiconducting single crystals. Advanced Materials. 2015;**27**:7213-7220

[28] Basiricò L, Ciavatti A, Cramer T, Cosseddu P, Bonfiglio A, Fraboni B. Direct X-ray photoconversion in flexible organic thin film devices operated below 1 V. Nature Communications. 2016;**7**:1-9

Section 5

The Applications of Microchips

Chapter 7

CMOS Integrated Circuits for Various Optical Applications

Sung Min Park

Abstract

This chapter presents several CMOS integrated circuits (ICs) realized for various optical applications such as high-definition multimedia interface (HDMI), light detection and ranging (LiDAR), and Gigabit Ethernet (GbE). First, 4-channel 10-Gb/s per channel optical transmitter and receiver array chipset implemented in a 0.13-μm CMOS process are introduced to realize a 10-m active optical cable for HDMI 2.1 specifications. Second, a 16-channel optical receiver array chip is realized in a 0.18-μm CMOS technology for LiDAR applications. Third, a 40-GHz voltage-mode mirrored-cascode transimpedance amplifier (MC-TIA) is implemented in a 65-nm CMOS for a feasible 100-GbE application. Even with advanced nano-CMOS technologies, we have suggested novel circuit techniques for optimum performance, such as input data detection (IDD) for low power, feedforward and asymmetric preemphasis for high speed, double-gain feedforward for high gain, selectable equalizer (SEQ) for specific bandwidth, mirrored-cascode for fully differential topology, etc. We believe that these novel circuit techniques help to achieve low-cost, low-power solutions for various optical applications.

Keywords: active optical cables, CMOS, GbE, HDMI, integrated circuits, LiDAR, TIA, VCSEL driver

1. Introduction

Optical fibers provide a number of advantages over copper-based electrical cables, which include wide bandwidth, low attenuation, low weight, low electromagnetic interference, low crosstalk between channels, etc. Particularly for high-speed digital interconnects, optical fibers may be the ultimate solution to achieve the desired performance. In this chapter, a few CMOS integrated circuits (ICs) are introduced for various optical applications such as high-definition multimedia interface (HDMI), light detection and ranging (LiDAR), and Gigabit Ethernet (GbE).

Section 2.1 presents 4-channel CMOS transmitter (Tx) and receiver (Rx) chipsets for the applications of 10-Gb/s per channel HDMI active optical cables (AOC). Section 2.2 describes a feedforward voltage-mode CMOS Rx IC for LiDAR applications. Section 2.3 introduces a 40-GHz CMOS Rx IC. Then, conclusion is followed.

IntechOpen

2. Circuit description

2.1 CMOS chipsets for HDMI active optical cables

High-performance networking and computing systems mandate high-speed optical interconnects to satisfy the extreme bandwidth requirements [1, 2]. Previously, parallel optical interconnects could provide terabit-per-second data bus in board scale [1] and also multi-gigabit-per-second data transport for mega-cloud systems, reaching a 100-m distance [2]. We have recently demonstrated active optical cables specified by HDMI 2.0 standard for true 4K video at 60-Hz resolution for 10-m distance, where 4-channel CMOS Tx and Rx chipsets were integrated on a printed circuit board (PCB) with pluggable connectors at its both ends to transport data via plastic optical fibers (POF) [3]. POF is well known for its benefits over costly glass optical fibers such as low cost, lightweight, resilient to bending, etc. [4]. In this section, we demonstrate a 10-m AOC utilizing graded-index POF with −60-dB/km loss characteristics that equips 4-channel CMOS transmitter and receiver chipsets to support HDMI 2.1 specification, i.e., true 8 Mpixel/60 fps display with no data encoding or compression. For this purpose, it is necessary to align optical devices, optical subassembly, and POF precisely within the tolerance range of ±10 μm.

Figure 1 shows the block diagram of the 4-channel optical ICs, where a 4-channel Tx and Rx chipsets are separately integrated with optical devices. Here, we have employed a number of circuit techniques to optimize the performance, which include feedforward preemphasis at Tx for high-speed operations; input data detection (IDD) for automatic turning off each vertical-cavity surface-emitting laser (VCSEL) diode during its idle time to lower current consumption; double-gain feedforward transimpedance amplifier (TIA) for high gain; selective equalizer for either 6 or 10 Gb/s, depending upon desired HDMI specification; and photodiode

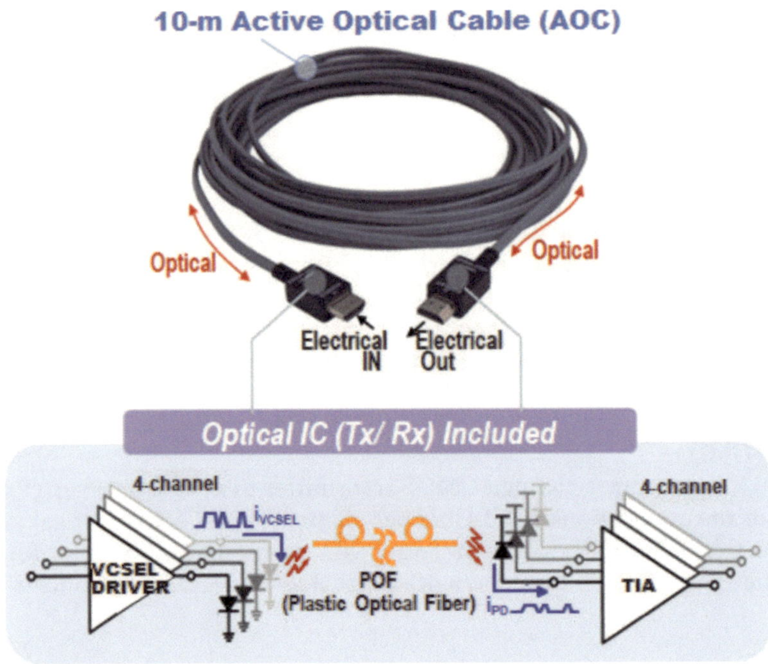

Figure 1.
Block diagram of the 4-channel optical ICs.

monitor for checking if each photodiode emits appropriate photocurrents to the 4-channel Rx array chip.

2.1.1 Optical power budget

There are various sources of coupling loss occurred in its optical alignment. For example, 3-dB coupling loss occurs at the interface of a VCSEL diode to prism due to 50% coupling efficiency, whereas 1-dB coupling loss occurs at the interface of a photodiode to prism [5]. The optimal pitch between VCSEL diodes and photodiodes was carefully selected to be 400 μm to prevent extra coupling loss from misalignment.

Meanwhile, the low-cost POF shows −60-dB/km attenuation, resulting in 1-dB loss. Also, the thermal loss of a VCSEL diode is typically 2 dB at 70°C. Hence, the optical power budget is set to 10 dB including 3-dB additional margin, which leads to the feasible assumption of 0-dBm Tx power and −10-dBm Rx sensitivity.

2.1.2 VCSEL driver

Figure 2 shows the schematic diagram of a 10-Gb/s VCSEL driver that consists of a main driver, a pre-driver, an EQ, and an input buffer. The main driver operates with two current sources, i.e., the bias current (I_{BIAS}) and the modulation current (I_{MOD}). When M_{5N} in the main driver is turned off, the current sum (I_{BIAS}+I_{MOD}) flows through the VCSEL diode. When M_{5N} is on, only I_{BIAS} is supplied to the VCSEL diode. The feedforward preemphasis is conducted by using a capacitor (C_{FF}) to alleviate the distortion effects of the output waveforms from the bond-wire inductance and the parasitic capacitance of a VCSEL diode. Simulations confirm 19.2% faster rising time in the output waveforms.

Considering the device reliability of VCSEL diodes, it is not clever to keep I_{BIAS} to flow continuously through the array chip because it will rise the device

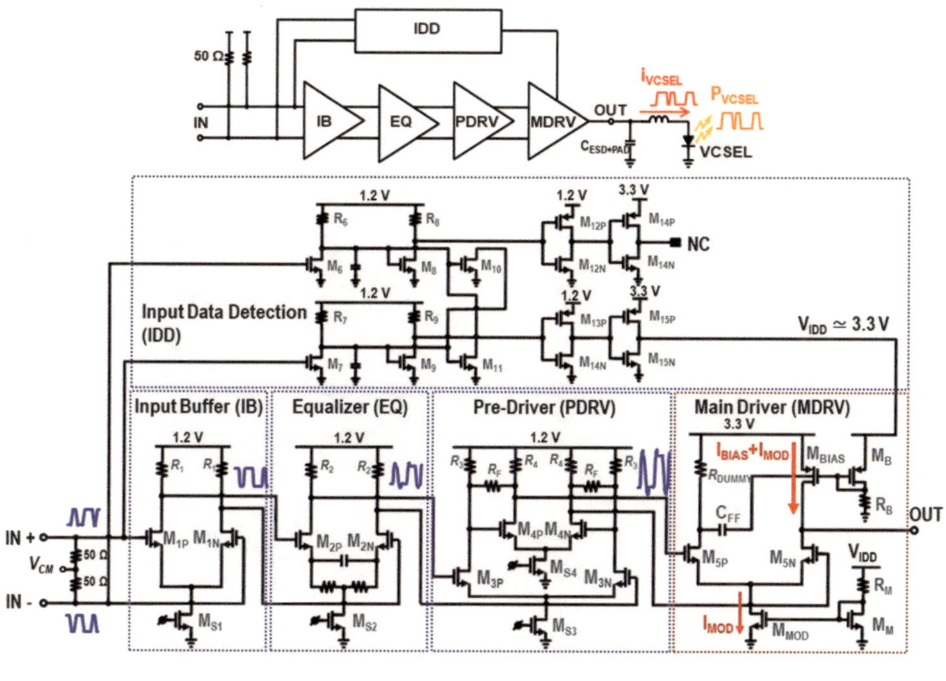

Figure 2.
Schematic diagram of the VCSEL driver.

temperature and thus the slope efficiency and expectant life period of VCSEL diodes will be severely deteriorated. Hence, input data detection circuit can be employed to avoid the superfluous current flow by turning off VCSEL diodes when no input signal transitions occur. Only with the emergence of input data, IDD detects the data transition and generates an average DC voltage, i.e., 3.3 V in this work through an active low-pass filter (LPF).

Then, this DC voltage turns on the two current sources in the main driver. Certainly, the turn-on delay of the current sources should be shorter than the signal delay from the main driver input.

Figure 3 shows the chip microphotograph of the 4-channel VCSEL driver array realized in a 0.13-μm CMOS process, where the chip core of each channel occupies the area of 350 × 250 μm^2. Each channel dissipates 21.25 mA (max.), in which the main driver consumes 11.6 mA.

Figure 4 demonstrates the optically measured output eye diagrams at 10 Gb/s with a 10-m POF connected to a digital communication analyzer (DCA Agilent 861150D), where it is clearly seen that there is no overlap area of 0.6-UI optical mask condition.

Table 1 compares the performance of the 4-channel Tx array chip with prior arts, in which only this work provides the measured optical magnitude amplitude (OMA) with a 10-m attached.

2.1.3 Voltage-mode CMOS feedforward TIA

Typically, the front end of an optical Rx comprises a photodiode and a TIA that converts the incoming current signals from the photodiode into output voltages. For optimum performance, TIA is usually required to achieve high transimpedance gain, wide bandwidth, low noise, and low power consumption.

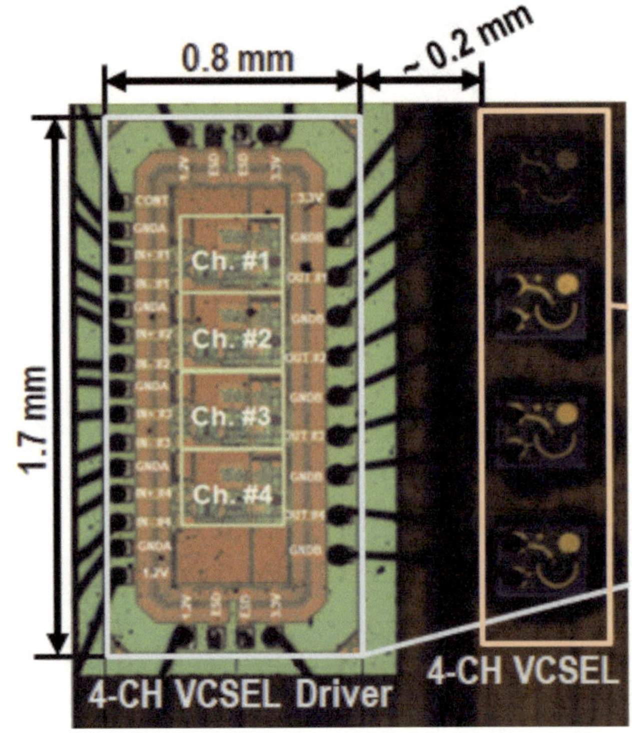

Figure 3.
Chip microphotograph of the 4-channel VCSEL driver array.

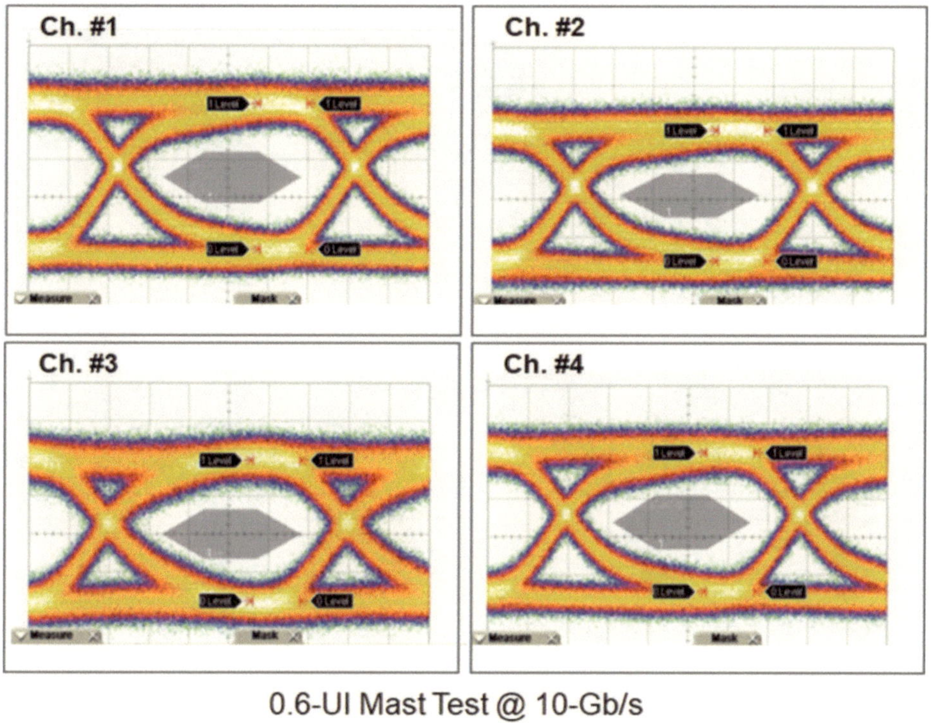

Figure 4.
4-channel eye mask at 10 Gb/s.

Parameters	[6]	[7]	[8]	[9]	[10]	This work
CMOS [nm]	65	65	90	28	65	**130**
Dara rate [Gb/s]	10	25	25	28	15	**10**
Channels	4	4	1	1	1	**4**
V_{DD} [V]	1.2/2.5	1.0/3.3	1.2/2.65	2.9	1.0/2.5	**1.2/3.3**
C_{VCSEL} [pF]	–	0.15	0.3	–	0.5	**0.85**
I_{MOD} [mA_{pp}]	4.0	6.7	–	5.5	6.0	**6.0**
I_{BIAS} [mA]	6.0	4–12	–	2.5	2.0	**5.6**
RMS jitter [% UI]	1.01	3.5	–	–	7.35	**4.8***
OMA [dBm]	–	−1	1.23	0.032	2.3	**>−6.77***
Current dissip. [mA]	16.3	67.6	38.6	17.6	23.8	**21.25**
Core area [mm^2]	0.125	0.353	0.006	0.001	0.04	**0.074**

**Measured with a 10-m POF attached.*

Table 1.
Performance summary of the 4-channel Tx array chip with prior arts.

To this end, we have developed a double-gain feedforward TIA, also known as the voltage-mode CMOS feedforward (VCF) TIA. **Figure 5** shows the schematic diagram that consists of the VCF input stage with DC offset current cancelation scheme, a single-to-differential converter, a selectable equalizer (EQ), an output buffer (OB), and a photodiode monitor. The double-gain feedforward input stage merges an inverter with a feedback resistor and a common-source amplifier

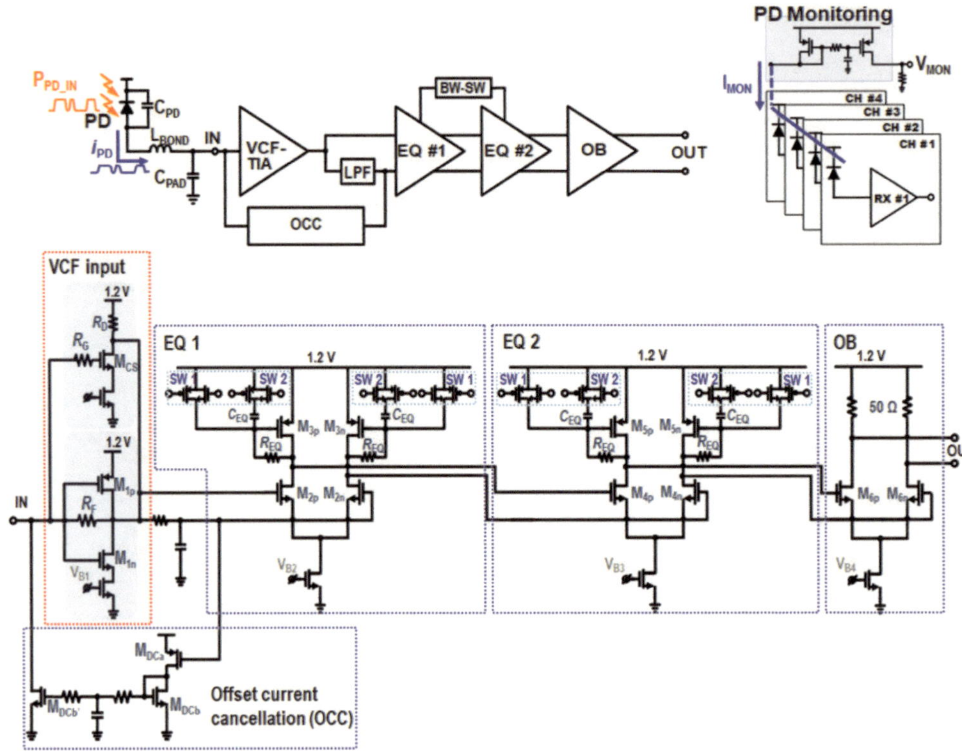

Figure 5.
Schematic diagram of the VCF-TIA.

together so that the transimpedance gain can be twice higher than a conventional inverter input stage. A two-stage EQ is followed not only to extend the bandwidth but also to select the operation speed to be either 6 Gb/s or 10 Gb/s with respect to the HDMI specification. With the switch 1 (SW1) turned on, the bandwidth is extended to 6 GHz for 10-Gb/s operations with a slight gain peaking. With the switch 2 (SW2) turned on, the bandwidth shrinks to 4 GHz for 6-Gb/s operations.

As for monitoring the input signal strength, the received signal strength indicator (RSSI) is utilized with an external resistor. However, a multichannel Rx array chip mandates the same number of external resistors for RSSI, which complicates the PCB assembly. Therefore, we suggest a simple PMOS current mirror circuit that generates an average DC voltage via only one fixed external resistor, thereby detecting the photodiode failure easily.

Figure 6 shows the chip microphotograph of the 4-channel Rx array chip realized in a 0.13-μm CMOS process, where the core of each channel occupies an area of 400 × 150 μm^2. Each channel consumes 21.2 mA, in which the OB dissipates 8.2 mA. For optical measurements, we have utilized a 4-channel GaAs p-i-n photodiode array that provides 0.6-A/W responsivity.

Figure 7 demonstrates the optically measured output eye diagrams of the 4-channel Rx array chip at the digital oscilloscope (Agilent DSA-X 92004A) via a 10-m POF, which was driven by a pulse pattern generator (Agilent ParBERT 81250) with 2^{31}-1 PRBS differential input swings of 800 mV_{pp} at different data rates of 5 Gb/s, 6.25 Gb/s, 8 Gb/s, and 10 Gb/s, respectively. When the SW1 was turned on, it obtained wide and clean eyes up to 10 Gb/s. Otherwise, the highest achievable data rate was 6.25 Gb/s, where the total jitter was measured to be 34.5 ps_{pp}. Also, the bit error rate (BER) of the 4-channel Rx array chip was measured by utilizing an error detector (Agilent ParBERT 81250-N4873A). With a 10-m POF attached, the

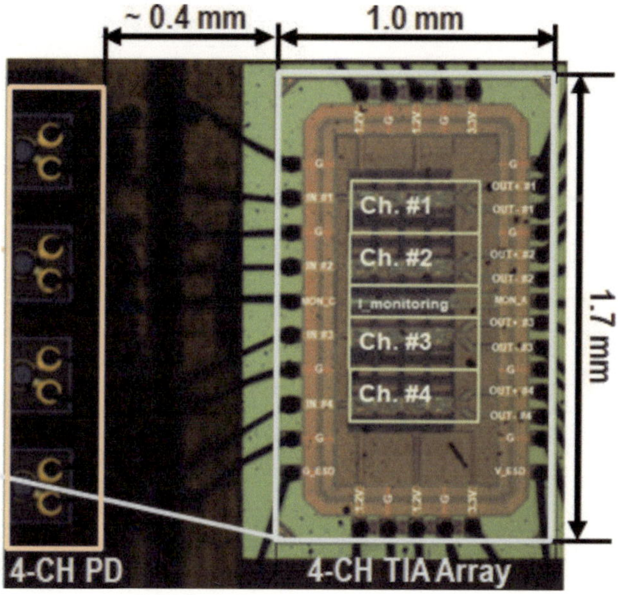

Figure 6.
Chip microphotograph of the 4-channel VCF-TIA array.

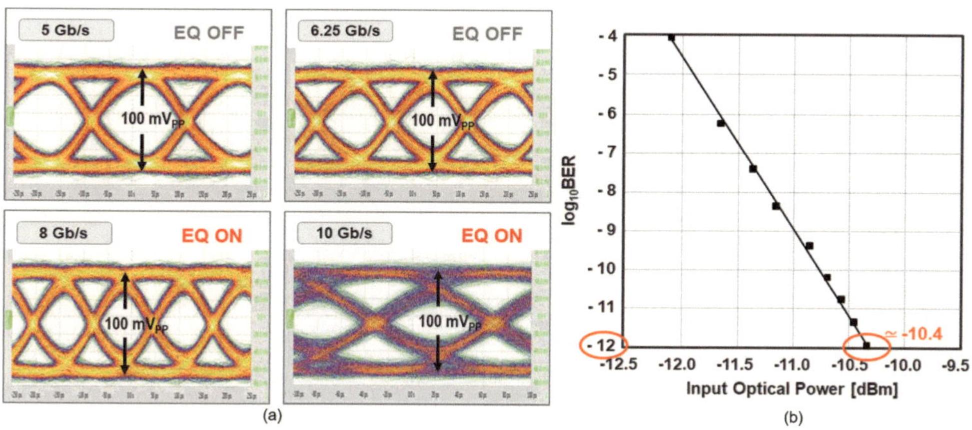

Figure 7.
Measured eye diagrams.

sensitivity of the 4-channel Rx array chip was measured to be -10.4 dBm for 10^{-12} BER at 10-Gb/s data rates.

Table 2 compares the performance of the 4-channel Rx array chip with prior arts, in which it should be noted that this work provides the measured optical sensitivity with a 10-m attached.

2.2 CMOS Rx IC for LiDAR

Light detection and ranging systems utilize laser pulses to detect surrounding targets efficiently and thus to characterize the scene in three-dimensional images [15, 16]. Therefore, LiDARs can be exploited to various applications such as unmanned vehicles to recognize pedestrians, driving lanes, natural objects, other vehicles, etc. For obtaining high-resolution images, a linear-mode LiDAR sensor with a multichannel optical Rx array can be an effective solution because the Rx

Parameters	[11]	[12]	[13]	[14]	This work
CMOS [nm]	130	180	65	65	**130**
Dara rate [Gb/s]	10	10	12	18	**10**
Channels	1	1	1	1	**4**
V_{DD} [V]	1.5	1.8	1.2	1.2	**1.2**
C_{PD} [pF]	0.25	0.2	0.05	Integrated	**0.25**
TZ gain [dBΩ]	50	68.3	–	102	**56.7**
BW [GHz]	7.0	7.0	3.5	12.5	**6.0**
Sensitivity for 10^{-12} BER [dBm]	−12.4**	−19	−16.8	−4.9	**−10.4***
Power dissip. [mW]	7.5^{Δ}	81	23	48	**25.4**
Energy efficiency [pJ/b]	0.75^{Δ}	8.1	1.92	2.7	**2.54**
Core area [mm^2]	0.016^{Δ}	0.78	0.12	0.23	**0.06**

**Measured with a 10-m POF attached.*

***Electrically estimated.*

$^{\Delta}$Single-ended TIA input stage only.

Table 2.
Performance summary of the 4-channel Rx array chip with prior arts.

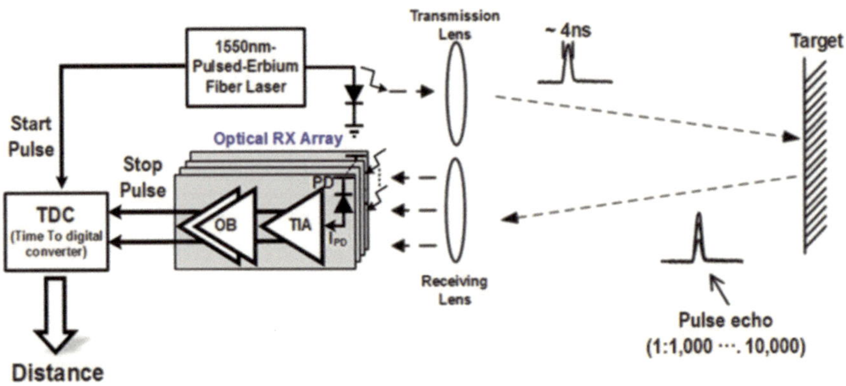

Figure 8.
Simplified block diagram of a typical LiDAR system.

array not only improves the object detection rate but also reduces the confusion matrix of point classification [17].

Figure 8 shows the simplified block diagram of the front-end circuitry in a typical linear-mode LiDAR sensor, which consists of a multichannel TIA array and a time-to-digital converter (TDC). As a photodetector, avalanche photodiodes (APDs) are exploited to acquire the detection range up to several tens of meters [18–21].

However, APDs need very high bias voltages of 50–200 V, thereby requiring overcurrent protection circuitry to avoid device saturation or damage. In this work, we have utilized an InGaAs p-i-n photodiode with 0.9-A/W responsivity biased with a low supply voltage of 5 V.

As a TIA, the most popular configuration has been a voltage-mode inverter with a feedback resistor. But, this inverter TIA has an inherent design trade-off between transimpedance gain and bandwidth, which may lead to considerable noise increase and sensitivity degradation [22]. Therefore, we have employed the

VCF-TIA in this work, hence clearly detecting the targets of 5% reflection rate within the range of 0.5–25 m.

2.2.1 Overview of linear-mode LiDAR sensor

The linear-mode LiDAR sensor emits optical laser pulses with 4-ns pulse width from a 1550-nm pulsed erbium fiber laser with the average power of 0.34–1.4 W and at the rate of 25 kHz. Also, a beam spread with the fan angle of 15o and the line intensity uniformity of less than 10% is utilized to transform laser beams into straight lines in far field. As a target, a $1 \times 1\ m^2$ black panel of which reflection rate is only 5% is utilized. In the Rx module, a 16-channel VCF-TIA array chip is wire bonded to a 16-channel InGaAs p-i-n photodiode array (Hamamatsu G7150-16) on an FR4 PC board.

2.2.2 VCF-TIA

TIA design for LiDAR sensors mandates wide-range transimpedance gain, i.e., high gain to detect the minimum input pulses for long-range detection versus low gain for short-range detection, wide bandwidth to recover the reflected narrow pulses, low noise for weak signal detection, and low power dissipation per channel to guarantee the reliability of multichannel Rx chips.

Figure 9 depicts the schematic diagram of the VCF-TIA, which comprises the VCF input stage for current-to-voltage conversion, a low-pass filter for single-to-differential conversion, a differential gain stage for gain boosting, and an OB for 50-Ω impedance matching. First, the small-signal analysis shows that the input resistance and the mid-band transimpedance gain (Z_T) of the VCF input stage are given by

$$R = \frac{v}{i} = \frac{\left(1 + \frac{R}{r_{o1}\|r_{o2}\|\|r_{o3}\|R_L}\right)}{\left(g_{m1} + g_{m2} + g_{m3} + \frac{1}{r_{o1}\|r_{o2}\|\|r_{o3}\|R_L}\right)} \cong \frac{1 + \left(3R_f/r_o\right)}{g_{m1} + g_{m2} + g_{m3}} \cong \frac{1}{g_{m1} + g_{m2} + g_{m3}} \quad (1)$$

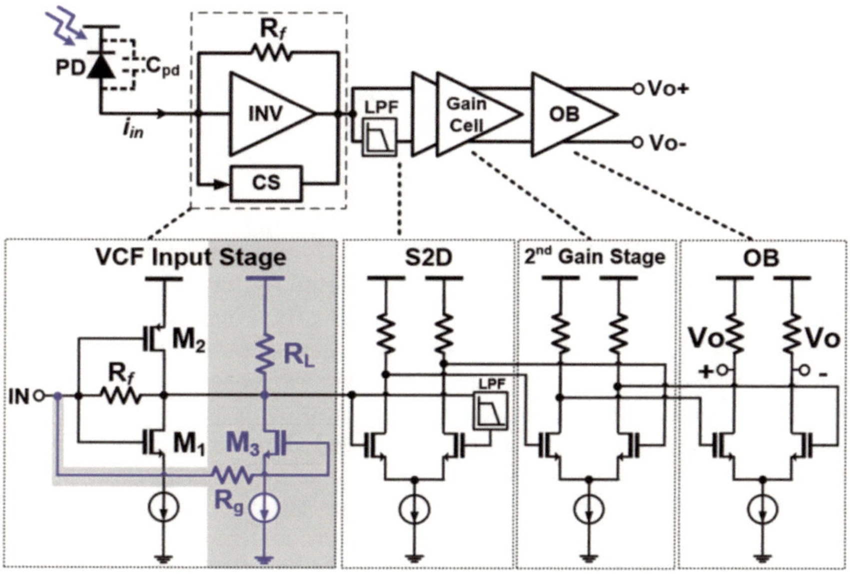

Figure 9.
Schematic diagram of the VCF-TIA.

$$Z_T = \frac{v_{out}}{i_{in}} = \frac{1 - R_f(g_{m1} + g_{m2} + g_{m3})}{\frac{1}{R_L} + (g_{m1} + g_{m2} + g_{m3})} \cong -R_f \tag{2}$$

where R_f is the feedback resistance, the load resistance (R_L) is assumed to be very large, and $g_{mi(i=1\sim3)}$ and $r_{oi(=1\sim3)}$ represent the transconductance and the output resistance of a transistor $M_{i(i=1\sim3)}$, respectively.

It is clearly seen that the input resistance (R_{in}) can be lowered by increasing g_m and thus R_f can be enlarged twice higher than in a conventional inverter TIA, so that the transimpedance gain can be doubled.

Second, the bandwidth (f_{-3dB}) of the VCF input stage is largely determined by the time constant (τ_3) at the gate of M_3 because the input time constant (τ_{in}) can be non-dominant by the boosted g_m and the output time constant (τ_{out}) can be negligible due to the small drain-bulk capacitances. Therefore, the bandwidth is given by

$$f_{-3dB} = \frac{1}{2\pi(\tau + \tau_3 + \tau_{out})} \cong \frac{1}{2\pi\tau_3} \frac{1}{2\pi R_g\{C_{gs3} + C_{gd3}[1 + (g_{m1} + g_{m2} + g_{m3})R_f]\}} \tag{3}$$

where R_g is the series resistance at the gate of M_3 and C_{gs3} and C_{gd3} represent the capacitances of the feedforward transistor (M_3).

Third, the input-referred equivalent noise current spectral density of the VCF input stage is given by

$$\begin{aligned} i_{eq}^{\prime 2} &\cong i_{Rf}^{\prime 2} + \left(\frac{1}{R_f} + sC_T\right) \times \left[\frac{i_{d1}^{\prime 2}}{g_{m1}^2} + \frac{i_{d2}^{\prime 2}}{g_{m2}^2} + \frac{\left(i_{d3}^{\prime 2} + i_{RL}^{\prime 2}\right)}{g_{m3}^2} + v_{Rg}^{\prime 2}\right] \\ &\cong \frac{4kT}{R_f} + 4kT\left[\Gamma\left(\frac{1}{g_{m1}} + \frac{1}{g_{m2}} + \frac{1}{g_{m3}}\right) + R_g\right] \times (\omega C_T)^2 \end{aligned} \tag{4}$$

where k is the Boltzmann's constant, T is the absolute temperature, Γ is the noise factor of a MOSFET, and i'_{Rf}, i'_{RL}, and $i'_{di(i=1\ 3)}$ represent the thermal noise current spectral densities of R_f, R_L, and $M_{i(i=1\sim3)}$, respectively. v'_{Rg} is the thermal noise voltage spectral density of R_g, and C_T ($=C_{pd} + C_{in}$) is the total input capacitance of the VCF input stage that consists of the photodiode capacitance (C_{pd}) and the input parasitic capacitance (C_{in}) of the VCF input stage. It is clearly seen that the critical factors to determine the high-frequency noise are R_f, g_m, and R_g. Increasing R_f boosts the transimpedance gain and lowers the low-frequency noise, while decreasing R_g not only reduces the high-frequency noise but also extends the bandwidth. The values of g_m should be judiciously selected to optimize the design trade-off between input resistance, bandwidth, and high-frequency noise.

Meanwhile, the VCF-TIA equips the function of overcurrent signal detection because the unexpected situations of overcurrent signals may often occur on real roads such as collisions with pedestrians and other cars. Therefore, all the 16 channels of the VCF-TIA array chip exploit automatic gain control (AGC) scheme to avoid the potential danger that consists of 4 NMOS switches with series resistors. This four-level gain control is satisfactory to accommodate the input photocurrents from 1 μA_{pp} to 1.1 mA_{pp}. All the switches are closed simultaneously with a control voltage (V_{cont}) larger than 0.65 V which corresponds to 800 μA_{pp} input currents. For a larger current, the VCF-TIA output would be saturated.

Test chips of the 16-channel VCF-TIA array were realized in a 0.18-μm CMOS process. **Figure 10** shows the test setup for the linear-mode LiDAR sensor, where an infrared camera was also used to capture black and white images for the target distance of 2, 5, 15, and 25 m, respectively.

The distance (R) between a target and the pulsed erbium fiber laser is estimated by

$$R = \frac{c \times t_{ns}}{2} = 0.15 \times t_{ns} \tag{5}$$

where c is the speed of light and t_{ns} is the round-trip time [15].

Figure 11 demonstrates the recovered output pulses with 340-mW average laser power, revealing that the Rx module can detect the target vividly within the range of 25 m.

Figure 12 shows the measured constant output pulses even with the variation of average optical powers from 0.65 W to 1.4 W, where the target is located at 0.5 m away from the laser source.

Figure 13 depicts the measured transimpedance gain, where the maximum value of 76 dBΩ is maintained for small input currents from the minimum detectable current of 1.14 μA_{pp} to the maximum current of 327 μA_{pp} with the photodiode responsivity of 0.9 A/W. Also, the minimum transimpedance gain becomes 41 times lower, i.e., 44 dBΩ. Hence, the input dynamic range (DR) defined by the ratio of the maximum and minimum detectable input currents times the gain variation, as described in [22], is given by

$$DR \equiv DR(highgainmode) \times max\ gain\ ratio = \left(\frac{327}{1.14}\right) \times (41) = 11,760 \tag{6}$$

2.3 High-speed CMOS receiver ICs

Recently, 100-Gigabit Ethernet (100 GbE) systems have received a great deal of attention [23]. Although quad 25-Gb/s per channel circuits can be a feasible solution in practice, there is still a need to increase the per channel bandwidth further so that a single-channel 100-Gb/s operation can be ultimately realized.

Previously, a number of high-speed TIAs have been introduced, in which the single-ended shunt-feedback topology was mostly preferred to achieve high sensitivity and low power consumption [24]. However, the single-ended circuit is vulnerable to common-mode noises occurred from power supply rails and silicon substrate, which might be detrimental in a 4-channel 25-Gb/s/ch optical Rx array chip for the applications of 100-GbE systems. Therefore, differential architecture is strongly desired even at the TIA input stage in order to reduce common-mode

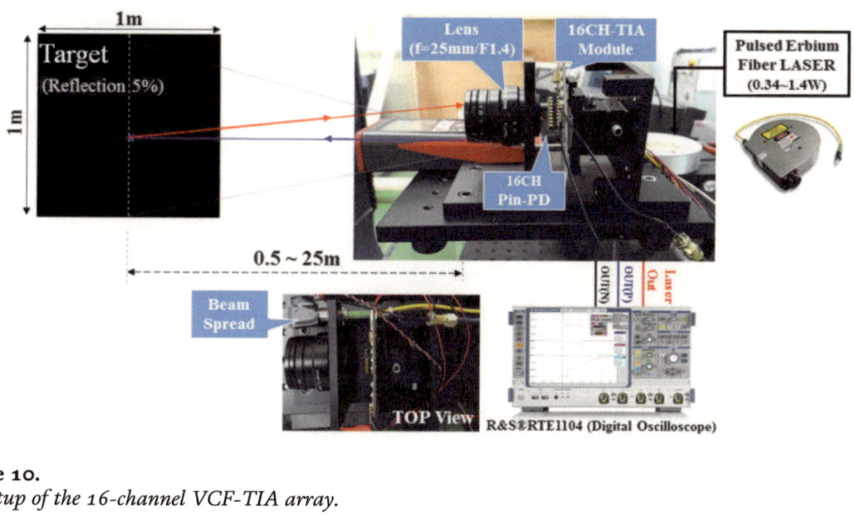

Figure 10.
Test setup of the 16-channel VCF-TIA array.

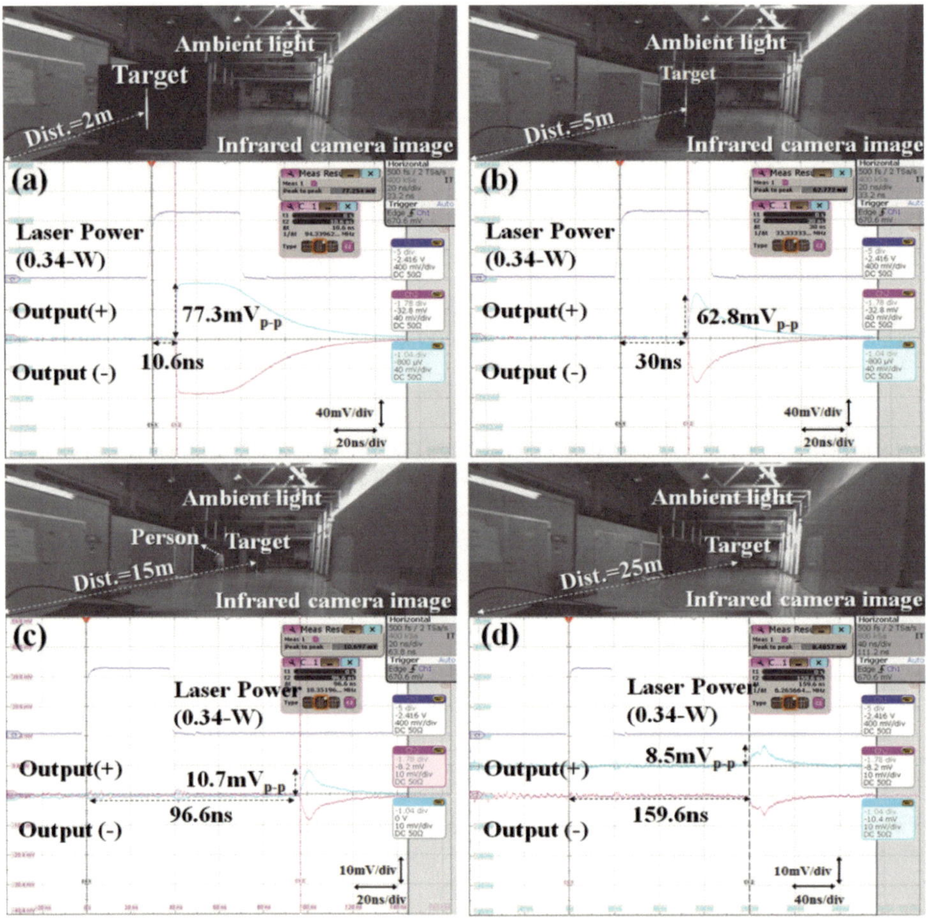

Figure 11.
Recovered output pulses from 2 to 25 m.

noises. In [25], two photodiodes were utilized to construct a fully differential TIA, which successfully alleviated the effects of the coupled common-mode noises. Yet, it was costly and rendered the PC board design complicated especially in the case of multichannel parallel interconnects. Alternatively, pseudo-differential structure was suggested, which however mandates either a passive low-pass filter (LPF) or a dummy TIA as a replica circuit [26]. The former cannot provide fully differential signaling without the following differential amplifier stages after the LPF, which certainly increases chip area and power consumption. The latter can hardly remove the DC offset between the TIA core and the dummy, let alone the increase of power consumption. Recently, we have presented a fully differential modified regulated cascode TIA in [27]. Yet, it suffered inherent noise degradation because of the current-mode common-gate input configuration.

In this chapter, a novel mirrored-cascode (MC) input configuration is introduced to overcome all these shortcomings, in which the NMOS cascode amplifier with a resistive feedback generates negative output voltages while its mirrored-cascode circuit via an AC-coupling capacitor yields positive counterparts. Since the MC input configuration shares the basic topology of a typical cascode TIA, it can provide an inherent advantage of noise performance over current-mode configurations such as a common-gate TIA [28], a regulated cascode TIA [29], and a current mirror TIA [30].

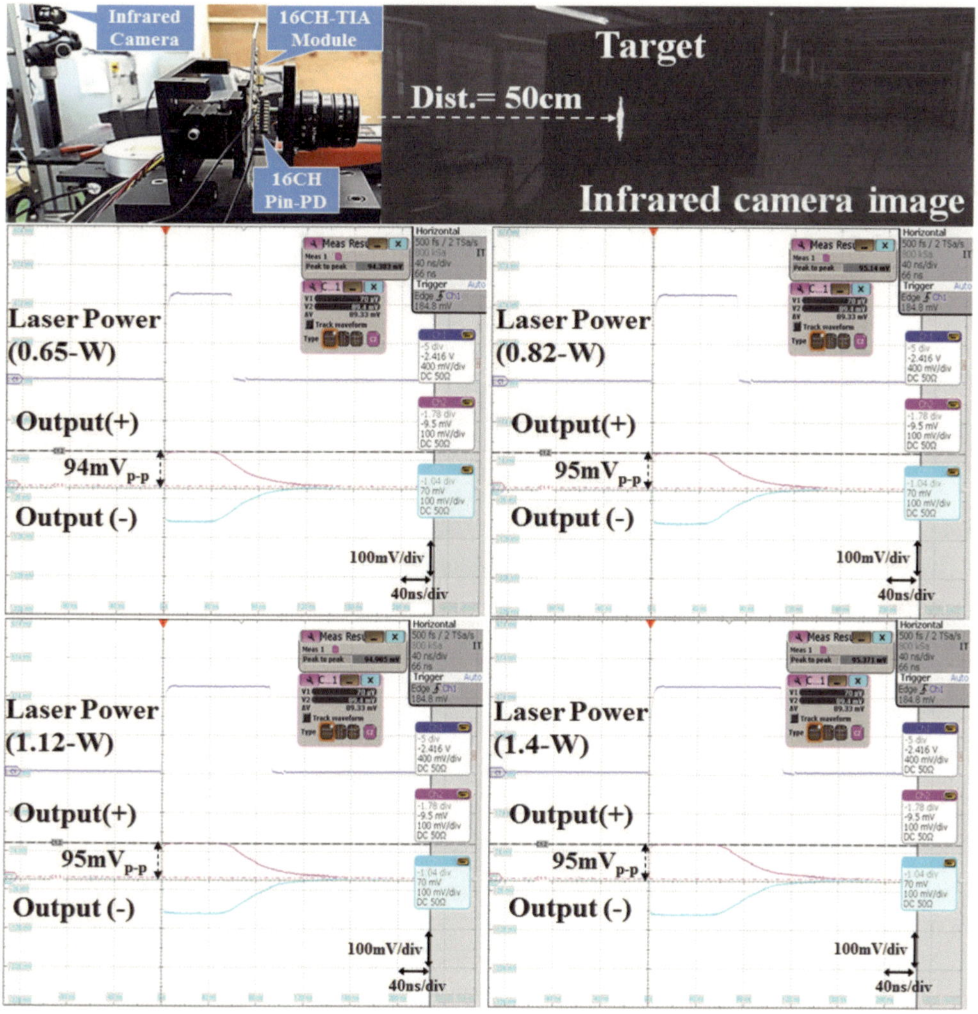

Figure 12.
Recovered output pulses within 0.5 m.

Meanwhile, a standard 65-nm CMOS process was utilized to implement the high-speed fully differential mirrored-cascode transimpedance amplifier (MC-TIA) with extensive exploitation of inductive peaking techniques to achieve 40-GHz bandwidth. In particular, asymmetric T-coil transformers were employed owing to their broadband characteristics, thereby reducing silicon area and lowering chip cost [31]. **Figure 14** shows the schematic diagram of the proposed MC-TIA.

2.3.1 Mirrored-cascode input configuration

Figure 15 depicts the simplified schematic diagram of the MC input stage, where the single-ended input current (i_{pd}) from a p-i-n photodiode flows into an NMOS cascode stage (M_1, M_3, R_1, and R_{F1}), giving rise to a negative output voltage (v_{o1}). It is noted that the drain voltage of M_1 is almost equal to the inverted gate voltage of M_1. Then, this negative voltage passes through an AC-coupling capacitor (C_c), appears at the gate of M_2 as an input signal of the mirrored-cascode stage (M_2, M_4, R_2, and R_{F2}), and hence generates a positive output voltage (v_{o2}) at the drain of M_4.

The input resistance of the MC input stage is given by

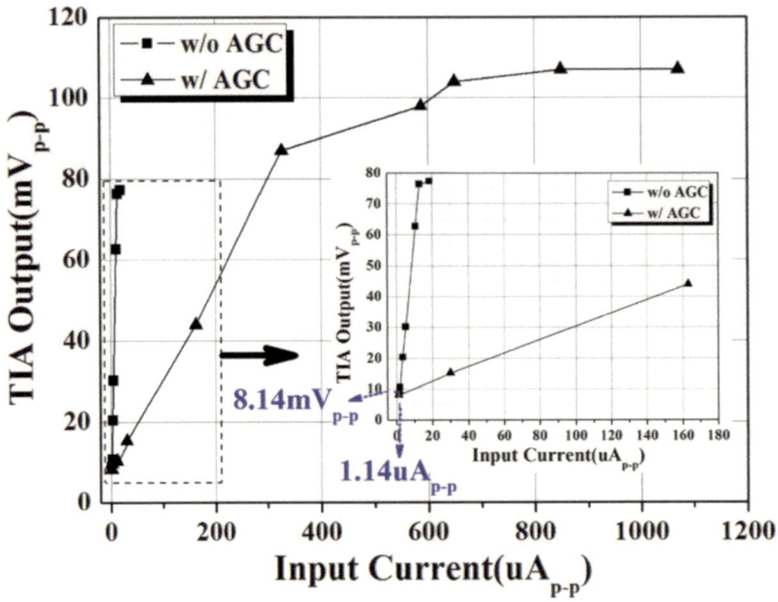

Figure 13.
Measured transimpedance gain with AGC.

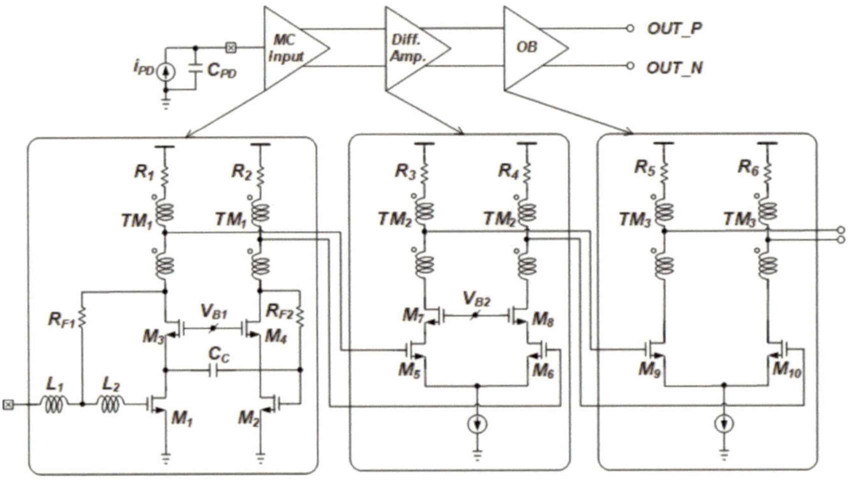

Figure 14.
Schematic diagram of the MC-TIA.

$$Z(0) = \frac{v}{i} = \frac{R_{F1}}{1 + g_{m1}R_1} \tag{7}$$

where g_{m1} is the transconductance of M_1.

The mid-band small-signal transimpedance gain at each drain node is given by

$$\begin{aligned} \frac{v_{o1}}{i_{pd}}(0) &= -\left(\frac{g_{m1}R_1R_{F1}}{1 + g_{m1}R_1}\right) \cong -R_{F1} \\ \frac{v_{o2}}{i_{pd}}(0) &= \frac{g_{m1}(g_{m2}R_2)R_{F1}}{(1 + g_{m1}R_1)g_{m3}} \cong R_{F1} \end{aligned} \tag{8}$$

where $g_{mi(i=1\sim3)}$ is the transconductance of $M_{i(i=1\sim3)}$.

The intrinsic output resistance (r_o) and the bulk transconductance (g_{mb}) of MOSFETs are omitted for simplicity. Provided that $g_{m2} = g_{m3}$, the transimpedance gain of both outputs would be the same as R_{F1}.

The equivalent noise current spectral density is given by

$$\begin{aligned} i_{eq}^{2'} &\cong i_{RF1}^{2'} + \frac{1}{g_{m1}^2}\left[\begin{array}{c} i_{d1}^{2'} + i_{R1}^{2'} + i_{RF2}^{2'} \\ \dfrac{+\left(i_{d2}^{2'} + i_{R2}^{2'}\right)}{g_{m2}^2 R_{F2}^2} \end{array}\right]\omega^2 (C_{PD} + C_1)^2 \\ &\frac{4kT}{R_{F1}} + \left[\begin{array}{c} \dfrac{4kT\Gamma}{g_{m1}} + \dfrac{4kT}{g_{m1}^2 R_1} + \dfrac{4kT}{g_{m1}^2 R_{F2}} \\ \dfrac{+4kT\Gamma}{g_{m1}^2 g_{m2} R_{F2}^2} + \dfrac{4kT}{g_{m1}^2 g_{m2}^2 R_{F2}^2 R_2} \end{array}\right]\omega^2 (C_{PD} + C_1)^2 \\ &\cong \frac{4kT}{R_{F1}} + \frac{4kT\Gamma}{g_{m1}}\omega^2 (C_{PD} + C_1)^2 \end{aligned} \tag{9}$$

where C_{in1} represents the parasitic capacitance of M_1, i.e., $C_{in1} = C_{gs1} + 2C_{gd1}$. Hence, it is clearly seen that the noise current spectral density of the MC input stage would be almost the same as that of an NMOS cascode input stage and certainly reduced by increasing g_{m2} ($=g_{m1}$).

2.3.2 Measured results

Test chips of the MC-TIA were fabricated in a standard 65-nm CMOS technology. **Figure 16** shows the chip microphotograph where the chip core occupies the

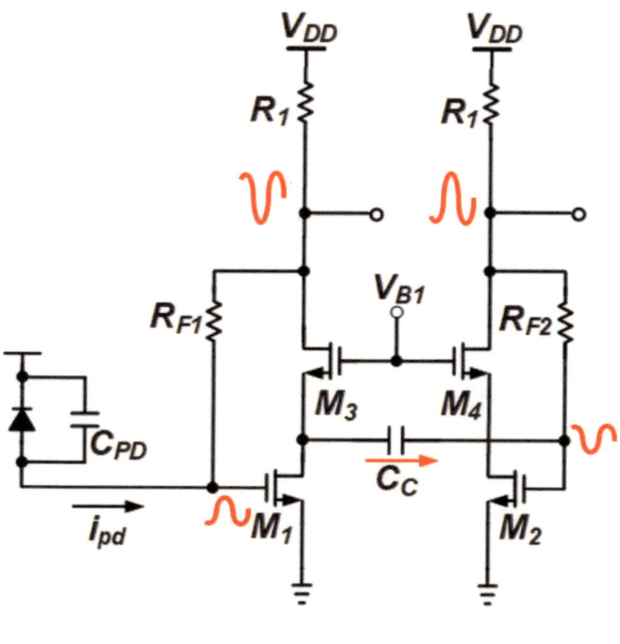

Figure 15.
Simplified schematic diagram of the VCF input stage.

area of 0.9 × 0.67 mm^2 including I/O pads. DC measurements reveal that the MC-TIA dissipates 55.2 mW (including the OB) from a single 1.2-V supply.

Figure 17 demonstrates the measured frequency response of the MC-TIA, where the differential transimpedance gain of 54 dBΩ and the −3-dB bandwidth of 40 GHz were measured with the gain flatness of ±1 dB. Also, the input impedance of the MC-TIA was measured to be in the range of 30–80 Ω within the bandwidth. The group delay variations of the MC-TIA were measured to be within ±10 ps. The single-ended integrated output noise voltage (1.42 mV_{rms}) of the MC-TIA was measured by using Agilent DCA 86100D oscilloscope in the absence of input signals [27, 29, 32].

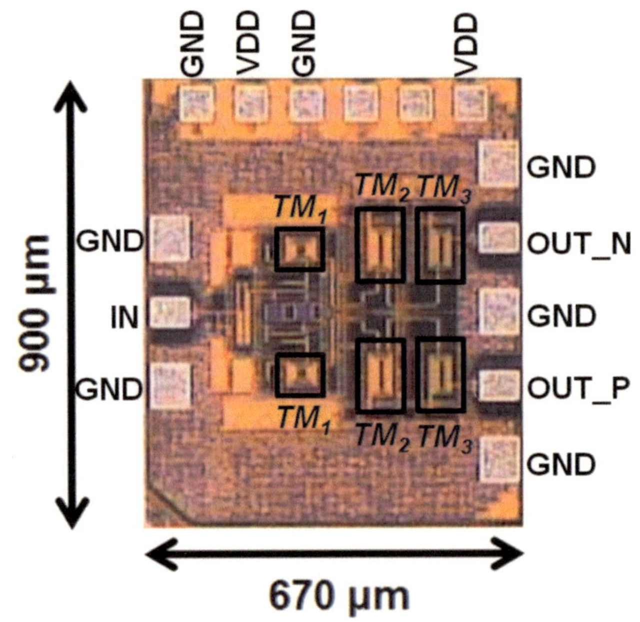

Figure 16.
Chip microphotograph of the MC-TIA.

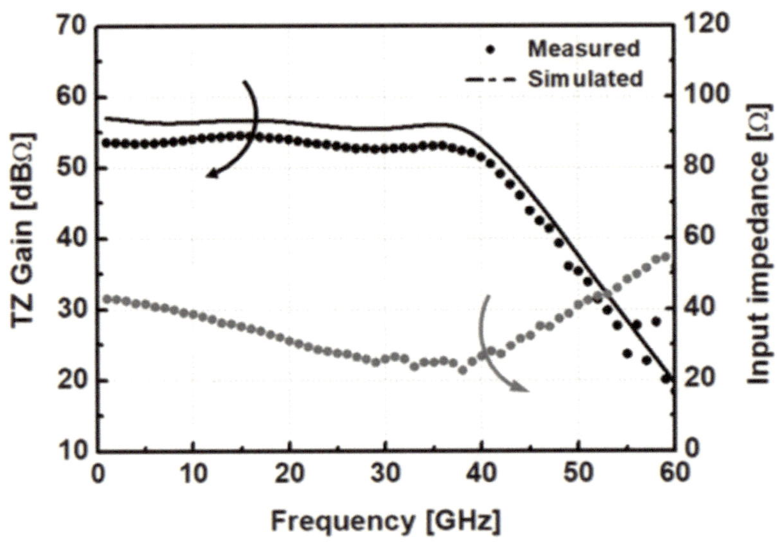

Figure 17.
Measured frequency response of the MC-TIA.

Considering the inherent oscilloscope noise of 1.01 mV_{rms}, the integrated input-referred noise current of the MC-TIA is given by

$$I_{n,in} = \frac{2\sqrt{(1.42\ \text{mV})^2 - (1.01\ \text{mV})^2}}{54\ \text{dB}\Omega} = 3.984\ \mu A_{rms} \tag{10}$$

Then, the average input-referred noise current spectral density is given by

$$I_{n,in,avg} = \frac{I_{n,in}}{\sqrt{BW}} = 19.9\ \text{pA}/\sqrt{\text{Hz}}, \tag{11}$$

which corresponds to the optical sensitivity of –13 dBm for 10^{-12} bit error rate with 0.6-A/W photodiode responsivity.

The eye diagrams of the MC-TIA were measured by utilizing RF probes and Anritsu MP1800A signal analyzer, of which operation speeds were limited to 32 Gb/s. It should be noted that the output voltage levels were measured with 50-Ω termination that caused 6-dB loss during the measurements.

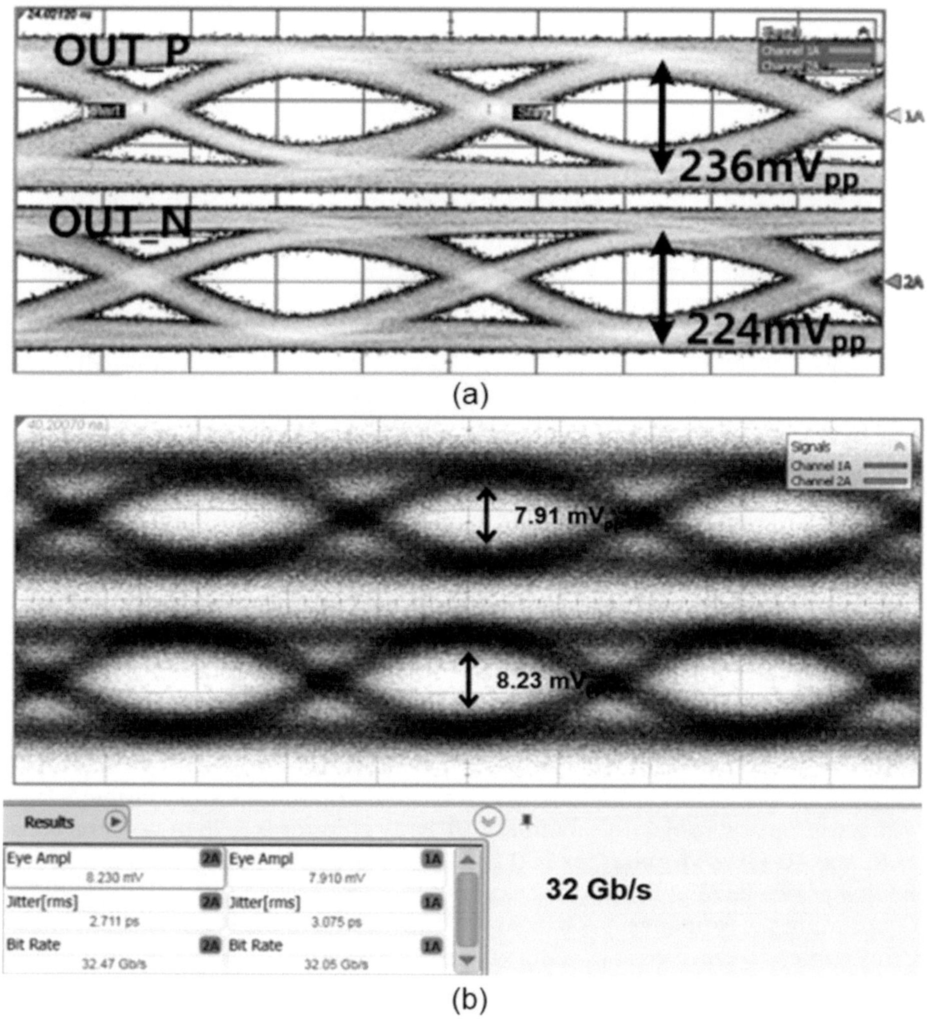

Figure 18.
Measured eye diagrams of the MC-TIA (a) with 1.5 mA_{pp} and (b) 100 μA_{pp} input currents, respectively.

Parameters	[24]	[34]	[33]	[32]	[30]	This work
CMOS [nm]	45 SOI	65	65	65	65	**65**
Architecture	INV (single)	SF (single)	RGC (single)	RGC+PA (single)	DMF (diff.)	**MC (diff.)**
V_{DD} [V]	1.0	1.6/2.2	1.2	1.0/3.3	1.2	**1.2**
BW [GHz]	30	40	21.6	21.4	50	**40**
PD cap. [fF]	60	40	200	N/A	50	**50**
TZ gain [dBΩ]	55	55	46.7	76.8	52	**54**
Noise current spectral density [pA/$\sqrt{\text{Hz}}$]	20.47	12.5	30	17.77	22.42	**19.8**
GD variation [ps]	±3.9	±32	N/A	N/A	N/A	**±10**
Power dissip. [mW]	9	122	39.9	137.5	49.2	**55.2**
Chip size [mm^2]	0.29	0.54	0.56	0.32	0.96	**0.6**

**INV, inverter; SF, source follower; RGC, regulated cascode; PA, post amplifier; DMF, dual-mode feedforward; single, single-ended; diff., differential; PD cap., photodiode capacitance.*
Bold: **Table 3** *summarizes the performance of the MC-TIA with the previously reported CMOS TIAs, showing a low-noise low-power fully differential solution for 100-GbE applications.*

Table 3.
Performance comparison with previously reported CMOS TIAs.

Figure 18(a) demonstrates the measured eye diagrams with 1.5 mA_{pp} 2^{15}-1 PRBS inputs at different data rates of 25 Gb/s and 32 Gb/s, respectively. It is clearly seen that the output voltage levels of larger than 210 mV_{pp} were measured with 50-Ω loads with the input currents of 1.5 mA_{pp}. The differential voltage swings of the MC-TIA were measured to be 236 mV_{pp} and 224 mV_{pp} (with less than 5.1% mismatch) at 25-Gb/s operations and 211 mV_{pp} and 198 mV_{pp} (with less than 6.2% mismatch) for 32-Gb/s operations. For both cases, delay mismatch at the differential outputs were measured to be less than 2 ps. **Figure 18(b)** demonstrates the measured eye diagrams with 100 μA_{pp} 2^{15}-1 PRBS inputs, where the voltage swings of 7.91 mV_{pp} and 8.23 mV_{pp} (with less than 4% mismatch) were achieved for differential outputs at 32-Gb/s operations. Even in this case, delay mismatch at the differential outputs was measured to be less than 2 ps.

Table 3 summarizes the performance of the MC-TIA with the previously reported CMOS TIAs.

3. Conclusions

We have demonstrated a number of CMOS integrated circuits for various optical applications, which included 4-channel 10-Gb/s/ch Tx and Rx array chipsets for HDMI active optical cables, 16-channel TIA array chip for 0.5–25 m range detection LiDAR, and 40-GHz TIA chip for 100 GbE. Even with advanced nano-CMOS technologies, we have proposed and exploited several novel circuit techniques for their optimum performance, such as input data detection for low power, feedforward and asymmetric preemphasis for high speed, double-gain feedforward for high gain, selectable equalizer for specific bandwidth, mirrored-cascode for fully differential topology, etc. We believe that the continuous introduction of novel circuit techniques is very crucial and necessary to develop low-cost CMOS ICs for various optical applications.

Author details

Sung Min Park
Department of Electronic and Electrical Engineering, Ewha Womans University, Seoul, Republic of Korea

*Address all correspondence to: smpark@ewha.ac.kr

References

[1] Nasu H et al. >1-Tb/s on-board optical engine for high-density optical interconnects. In: Proceedings of the IEEE Optical Fiber Conference (OFC'17); March 2017; USA. 2017

[2] Kuchta DM et al. 64 Gbps transmission over 57 m MMF using an NRZ modulated 950-nm VCSEL. In: Proceedings of the IEEE Optical Fiber Conference (OFC'17); March 2014; USA. 2014. pp. 1-3

[3] Hong C, Kim SH, Cha S, Park SM. A 10-meter active optical cable utilizing POF with 4x10-Gb/s CMOS transceiver chipsets. IEEE Photonics Journal. April 2019;**11**:6601011

[4] Polishuk P. Plastic optical fibers branch out. In: IEEE Communication Magazine. 2006. pp. 140-148

[5] Lee H–S et al. Ribbon plastic optical fiber linked optical transmitter and receiver modules featuring a high alignment tolerance. Optical Express. 2011:4301-4309

[6] Zeng Z et al. A compact low-power driver array for VCSELs in 65-nm CMOS technology. IEEE Transactions on Nuclear Science. 2017: 1599-1604

[7] Toru Y et al. 25-Gbps x4 optical transmitter with adjustable asymmetric pre-emphasis in 65-nm CMOS. In: Proceedings of the IEEE International Symposium on Circuits and Systems (ISCAS'14). June 2014. pp. 2692-2695

[8] Belfiore C et al. Common-cathode VCSEL driver in 90nm CMOS enabling 25 Gbit/s optical connection using 14 Gbit/s 850 nm VCSEL. Electronics Letters. 2015:349-351

[9] Szilagyi L et al. 30 Gbit/s 1.7 pJ/bit common-cathode tunable 850-nm VCSEL driver in 28 nm digital CMOS. In: Proceedings of the IEEE Optical Interconnects Conference (OIC'17). June 2017. pp. 51-52

[10] Kozlov V, Carusone AC. Capacitively coupled CMOS VCSEL driver circuits. IEEE Journal of Solid-State Circuits. September 2016;**51**:2077-2090

[11] Taghavi MH et al. 10-Gb/s 0.13-μm CMOS inductorless modified-RGC transimpedance amplifier. IEEE Transactions on Circuits and Systems I: Regular. 2015:1971-1980

[12] Li D, Liu M, Geng L. A 10-Gb/s optical receiver with sub-microampere input referred noise. IEEE Photonics Technology Letters. 2017:2268-2271

[13] Ahmed MG et al. A 12-Gb/s −16.8-dBm OMS sensitivity 23-mW optical receiver in 65-nm CMOS. IEEE Journal of Solid-State Circuits. 2018:445-457

[14] Pan Q et al. An 18-Gb/s fully integrated optical receiver with adaptive cascaded equalizer. IEEE Journal of Selected Topics in Quantum Electronics. 2016:6100509

[15] Hong C, Kim SH, Kim JH, Park SM. A linear-mode LiDAR sensor using a multi-channel CMOS transimpedance amplifier array. IEEE Sensors Journal. September 2018;**18**:7032-7040

[16] Ngo TH et al. Wideband receiver for a three-dimensional ranging LADAR system. IEEE Transactions on Circuits and Systems Brief II. 2013:448-456

[17] Lee MG, Baeg SH. Advanced compact 3D lidar using a high speed fiber coupled pulsed laser diode and a high accuracy timing discrimination readout circuit. In: Proceedings of SPIE Laser Radar Technology and Applications. 2012. pp. 1-12

[18] Nissinen J, Nissinen I, Kostamovaara J. Integrated receiver including both receiver channel and TDC for a pulsed time-of-flight laser rangefinder with cm-level accuracy. IEEE Journal of Solid-State Circuits. 2009:1486-1497

[19] Cho HS, Kim C-H, Lee S-G. A high-sensitivity and low-walk error LADAR receiver for military applications. IEEE Transactions on Circuits and Systems Regular I. 2014:3007-3015

[20] Kurtti S, Kostamovaara J. An integrated laser radar receiver channel utilizing a time-domain walk error compensation scheme. IEEE Transactions on Instrumentation Measurements. January 2011:146-157

[21] Brandl P et al. Optical wireless APD receiver with high background light immunity for increased communication distances. IEEE Journal of Solid-State Circuits. July 2016:1663-1673

[22] Zheng H et al. A linear dynamic range receiver with timing discrimination for pulsed TOF imaging LADAR applications. IEEE Transactions on Instrumentation Measurements. April 2018:1-8

[23] Kim SG et al. 40 GHz mirrored-cascode differential transimpedance amplifier in 65-nm CMOS. IEEE Journal of Solid-State Circuits. 2019:1468-1474

[24] Kim J, Buckwalter JF. A 40-Gb/s optical transceiver front-end in 45nm SOI CMOS. IEEE Journal of Solid-State Circuits. March 2012: 615-626

[25] Awny A et al. A dual 64Gbaud 10 kΩ 5% THD linear differential transimpedance amplifier with automatic gain control in 0.13 μm BiCMOS technology for optical fiber coherent receivers. In: Technical Digest of IEEE International Solid-State Circuits Conference (ISSCC'16), February 2016; San Francisco. USA: IEEE; 2016. pp. 406-407

[26] Chen W–Z, Cheng YL, Lin DS. 1.8 V, 10 Gb/s fully integrated CMOS optical receiver analog front-end. IEEE Journal of Solid-State Circuits. June 2005:1388-1396

[27] Kim SG et al. A 50-Gb/s differential transimpedance amplifier in 65 nm CMOS technology. In: Technical Digest of the IEEE Asian Solid-State Circuits Conference (A-SSCC'14). November 2014. pp. 357-360

[28] Han J et al. A 20-Gb/s transformer-based current-mode optical receiver in 0.13-μm CMOS. IEEE Transactions on Circuits and Systems Brief II. May 2010: 348-352

[29] Li C, Palermo S. A low-power 26-GHz transformer-based regulated cascode SiGe BiCMOS transimpedance amplifier. IEEE Journal of Solid-State Circuits. May 2013:1264-1275

[30] Yun JS et al. A 4Gb/s current-mode optical transceiver in 0.18 μm CMOS. In: Technical Digest of the IEEE International Solid-State Circuits Conference (ISSCC'09), February 2009; San Francisco. USA: IEEE; 2009. pp. 102-103

[31] Jin J–D, Hsu SSH. A miniaturized 70-GHz broadband amplifier in 0.13-μm CMOS technology. IEEE Transactions on Microwave Theory and Techniques. December 2008:3086-3092

[32] Takemoto T et al. A 4 × 25-to-28 Gb/s 4.9 mW/Gb/s -9.7dBm high sensitivity optical receiver based on 65nm CMOS for board-to-board interconnects. In: Technical Digest of the IEEE International Solid-State Circuits Conference (ISSCC'13), February. San Francisco; USA. 2013. pp. 116-117

[33] Bashiri S, Plett C, Aguirre J, Schvan P. A 40 Gb/s transimpedance

amplifier in 65 nm CMOS. In: Proceedings of the IEEE International Symposium on Circuits and Systems (ISCAS'14). May 2010. pp. 757-760

[34] Ding R, Xuan Z, Baehr-Jones T, Hochberg M. A 40-GHz bandwidth transimpedance amplifier with adjustable gain-peaking in 65-nm CMOS. In: Proceedings of the IEEE MidWest Circuits and Systems (MWCAS'14). pp. 965-968

Chapter 8

Area-Efficient Spin-Orbit Torque Magnetic Random-Access Memory

Karim Ali

Abstract

Spin-orbit torque magnetic random-access memory (SOT-MRAM) has shown promising potential to realize reliable, high-speed and energy-efficient on-chip memory. However, conventional SOT-MRAM requires two access transistors per cell. This limits the use of conventional SOT-MRAM in high-density memories. Thus, various architectures in the literature have been proposed to improve the area efficiency of the SOT-MRAM. In this chapter, these proposals are divided into two categories: non-diode-based SOT-MRAM and diode-based SOT-MRAM cells. The non-diode-based proposals may result in a 1-bit effective area saving up to 50% compared to the conventional SOT-MRAM, whereas the diode-based designs may result in 1-bit effective area-saving of up to 75%. However, the area saving may be accompanied by higher energy and reliability issue penalties. Therefore, here, the various proposals in the literature are presented, highlighting the pros and cons of each design. Moreover, the technology requirements to realize these proposals are discussed. Finally, the various designs are evaluated from both cell and system level perspectives.

Keywords: SOT-MRAM, MOM diode, SLC, MLC, MTJ

1. Introduction

In the era of the Internet of things (IoT), ultra-low power and energy-efficient computing become essential. The expected large number of the IoT nodes forces the limitation of their power budget [1]. In particular, for the majority of the IoT nodes that are powered by energy harvesters, their anticipated power budget can be as small as sub μWatt [1]. Thus, employing nonvolatile (NV) memory in these nodes would aid in reducing their power consumption. This is because leakage power constitutes a significant percentage of the total power consumed due to the long idle durations. Hence, the memory can be power gated, thanks to its nonvolatility, and consequently eliminates its leakage contribution. In addition, IoT nodes powered by energy harvesters may suffer from multiple durations of power discontinuities as a result of using the unreliable power source. Therefore, having NV memory aids to continue the computation from where it's stopped without restarting every single operation once the power goes down [2], which aids in increasing the overall performance as well. Furthermore, the different modules on these nodes need to be area efficient. Area efficiency aids in decreasing the overall parasitic contribution, which enhances the performance and energy efficiency. More importantly, the smaller the silicon area consumed by the IoT node, the lower the overall cost. The cost is an important metric for the success of a specific IoT design as it is targeted to have these IoT chips with prices as low as 50 cents. Thus, a combination of improved technologies and circuits is needed to achieve energy- and area-efficient designs.

Emerging devices such as a magnetic tunnel junction (MTJ) may be used to implement a nonvolatile memory, which would be called magnetic random-access memory (MRAM). An MTJ, highlighted in **Figure 1**, comprises two ferromagnetic (FM) layers separated by a tunneling oxide barrier. One FM layer has a pinned magnetization (i.e. pinned layer (PL)), while the other has a free magnetization (M_F) (i.e. free layer (FL)). Manipulating the direction of the FL to be either parallel (P) or anti-parallel (AP) to the PL magnetization direction determines the electrical resistance state of the MTJ to be either low resistance (R_P) or high resistance (R_{AP}), respectively. An MTJ offers different benefits like nonvolatility, programmability, high endurance, long state retention time and compatibility with CMOS fabrication flow. Consequently, the MRAM inherits the MTJ advantages, which makes MRAM a standout solution to implement a nonvolatile memory that is suitable to replace or co-exist with these leaky charge-based memories in both on-chip and off-chip memory.

The MRAM can be classified based on the writing technique employed to switch the MTJ resistance state. These techniques, such as field-induced magnetization reversal (FIM) [3], spin-transfer torque (STT) [4, 5], voltage-controlled magnetic anisotropy (VCMA) [6] and spin-orbit torque (SOT) [7–11], result in the different corresponding MRAM types like FIM MRAM, STT MRAM, VCMA MRAM, and SOT MRAM, respectively. FIM requires applying an external magnetic field to program the MTJ, which hinders its scalability and increases its energy consumption, while STT, VCMA, and SOT are known to be highly scalable [9]. STT programming has a common current path for both writing and reading the MTJ, whereby the write current (I_{write}) must flow through the MTJ directly. This results in higher writing voltage requirements, due to the MTJ high resistance, and device reliability degradation. Similarly, VCMA demands voltage application across the MTJ, which subjects it to reliability issues such as tunnel barrier breakdown. The FIM, STT, and VCMA technologies use the MTJ device independently in its two-terminal form, shown in **Figure 1(a)**. On the other hand, the SOT technology requires placing the MTJ over a heavy metal (HM) electrode from the FL side, which results in a three-terminal device as depicted in **Figure 1(b)**. In SOT technology, a charge current passes through the HM electrode that in return results in spin accumulation in the FL of the MTJ due to the spin-orbit interaction, which assists in the reversal of the FL magnetization to either P or AP state depending on the current direction relative to the FL easy axis. Thus, SOT programming solves

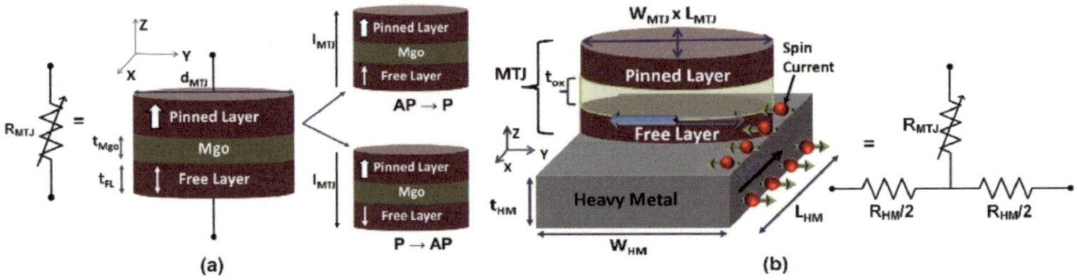

Figure 1.
(a) Two-terminal MTJ with two magnetization configurations reflects two resistance states depending on the write current direction if STT technology is used. (b) Three terminal SOT-MTJ.

the aforementioned issues [9]. Firstly, it features high energy efficiency due to its low critical current requirements and fast switching speed. In particular, due to the SOT-MRAM high switching speed (i.e. high performance) that can be down to 100 s of ps [12, 13], in addition to its nonvolatility and smaller cell area, it becomes more appealing to replace SRAM in cache memory applications. Secondly, the I_{write} flows through the low resistance heavy metal (HM) rather than the MTJ. This improves the device reliability, permits the usage of smaller write voltages, and increases the voltage headroom margin for the write transistor, which relaxes the current source design. Finally, the write and read paths are separated, which leads to a more optimized design as it permits satisfying the contradictory requirements for the access transistors sizing in both read and write modes. In the write mode, larger access transistor is required to supply higher I_{write} (at least larger than SOT-MTJ critical switching current), whereas in the read mode, smaller access transistor is required to reduce the read current to avoid read disturbances. Consequently, SOT-MRAM is considered as the most promising MRAM to realize reliable, high speed, and energy-efficient on-chip memory [14, 15].

However, the separation of the read and write paths comes with a penalty of requiring two access transistors per cell to fully isolate both paths of the nonselected cells in the conventional approach of implementing the SOT-MRAM, presented in **Figure 2(a)**. This increases the 1-bit effective area of the conventional SOT-MRAM, which limits its use in high-density memories. Thus, various SOT-MRAM cell designs in the literature have been proposed to improve the area efficiency of the SOT-MRAM. In this chapter, these proposals in the literature are going to be presented, highlighting the advantages and disadvantages of each design. The various proposals are divided into two main categories, which are diode-based and non-diode-based SOT-MRAM. Moreover, the technology requirements to realize these proposals are discussed. Finally, the proposals are evaluated from both cell and system level perspectives.

The chapter is organized as follows: Section 2 presented the nondiode-based SOT-MRAM proposals discussing their operation, pros, and cons. Similarly, the operation, pros, and cons of the various diode-based SOT-MRAMs are illustrated in Section 3. Section 4 evaluates the various diode and non-diode-based SOT-MRAM proposals from both cell and system-level perspectives. Finally, Section 5 provides the concluding remarks.

2. Nondiode-based SOT-MRAM

In this section, the various SOT-MRAM proposals in the literature that do not require employing a diode (or selector) in the cell are presented. Nondiode based SOT-MRAM cells only rely on access transistors for current flow control. Avoiding using diodes offer an advantage of a simpler fabrication process and may achieve higher energy efficiency as lower read voltages may be applied. The nondiode-based SOT-MRAMs in the literature include the conventional single-level cell (SLC) SOT-MRAM, as shown in **Figure 2(a)**, and two multi-level cell (MLC) proposals depicted in **Figure 2(c)** and **(d)**.

2.1 Conventional SLC SOT-MRAM

The conventional SOT-MRAM, depicted in **Figure 2(a)**, as discussed in various works [14, 15] requires two transistors to access one bit per cell. One of the transistors is used to control the write current flow (i.e. write Tx shown in **Figure 2(a)**) and the second transistor is required to control the read current flow (i.e. read

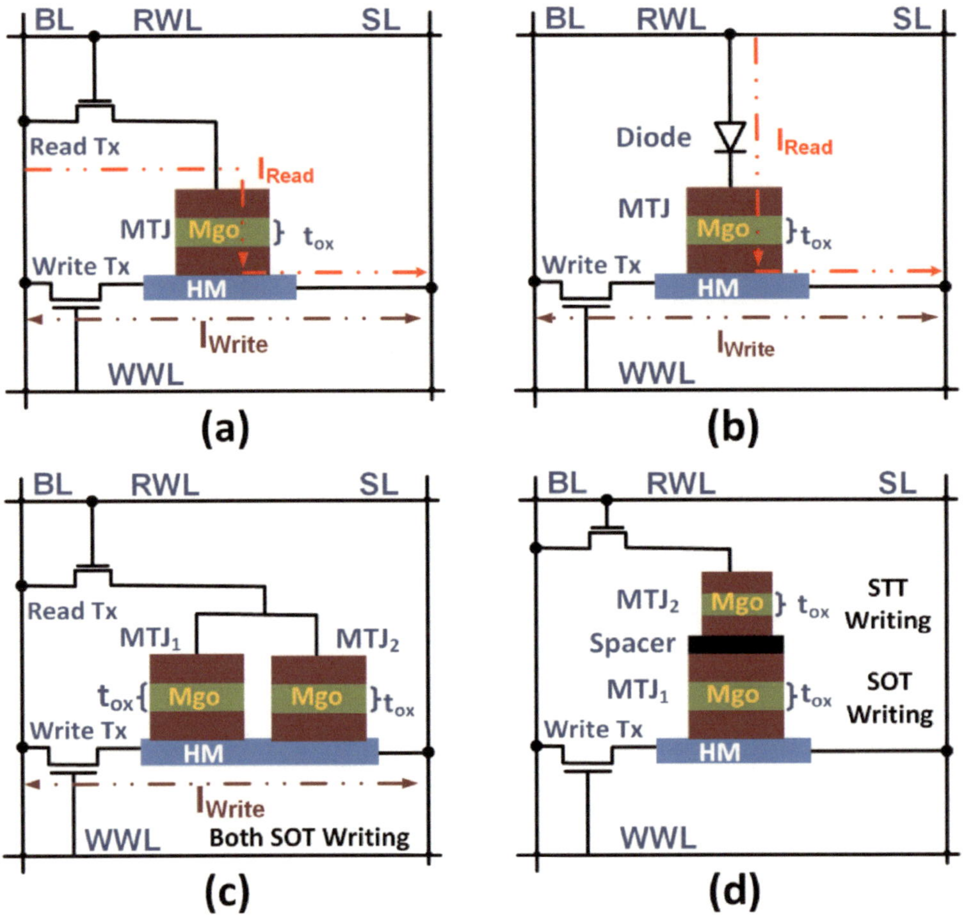

Figure 2.
Various SOT-MRAM proposals in the literature. (a) Conventional SOT-MRAM [15], (b) Schottky diode-based SLC SOT-MRAM [14], (c) P-MLC [15], and (d) S-MLC [15].

transistor). The two transistors are essential to fully isolate the nonselected bits to avoid nonintentional read and write of these bits when not selected. Furthermore, the existence of a single MTJ per heavy metal (HM) electrode to be programmed for each cell write operation maintains the high energy efficiency of SOT technology. In addition, the single-level cell sensing scheme for this cell enhances the distinguishability, bit error rate (BER), and read energy consumption. Consequently, the conventional SLC SOT-MRAM design, shown in **Figure 2(a)**, represents the baseline of the energy consumption for the SOT-MRAM. However, the usage of two transistors to access a single bit increases the 1-bit effective area significantly, which is estimated to be $69F^2$ using the design rules in [16] that is used in the area estimation throughout this chapter. This limits the application of the conventional SOT-MRAM as it will not be suitable for high-density memory applications. Thus, further SOT-MRAM cells are proposed in the literature to reduce the 1-bit effective area with an expected penalty in the various performance and energy metrics of the SOT-MRAM as discussed in the following sections.

2.2 MLC SOT-MRAM

Kim et al. [15] proposed to employ a multi-level cell (MLC) SOT-MRAM instead of the single-level cell (SLC) SOT-MRAM to reduce the 1-bit effective area. In the

MLC memory, each cell comprises two bits (i.e. two MTJs exists per cell) instead of only 1-bit (1 MTJ) in the SLC. This means that the two MTJ resistance combinations should represent four different resistance states instead of only two resistance states in the SLC. In this MLC design, the two bits are accessed by two transistors (i.e. effectively one transistor per bit), which results in an approximately 50% smaller 1-bit effective area compared to conventional SLC SOT-MRAM. However, there is still a margin to improve the 1-bit effective area than their estimated value of $34.5F^2$, using the design rules in [16], as illustrated later in Section 3. Moreover, their two proposed MLC designs, which are known as parallel MLC (P-MLC), shown in **Figure 2(c)**, and series MLC (S-MLC), depicted in **Figure 2(d)**, do suffer from various drawbacks that result in higher energy consumption and reliability issues as discussed below.

2.2.1 Parallel multi-level cell SOT-MRAM (P-MLC SOT-MRAM)

P-MLC, depicted in **Figure 2(c)**, encloses two MTJs in-parallel and both placed side-by-side over a common HM electrode. The main advantage of this cell structure is that the two bits are accessed by two transistors, which results in approximately 50% reduction in the 1-bit effective area. As the two MTJs are placed on the HM, both of the MTJs may be programmed by SOT effect. Writing the two MTJs with identical bits ('00' or '11') can be done using only one write pulse with duration following the slower MTJ. However, to write independent bits on the two MTJs ('01' or '10'), the two MTJs must have different critical currents (I_c) (i.e. the MTJs with smaller I_c switches faster for the same supplied current amplitude). Thereafter, either time-dependent or current-dependent writing [17] can be adopted. In time-dependent writing, a constant write current flows through the HM electrode, where both MTJs are firstly programmed within pulse duration t1, followed by the programming of faster MTJ2 with time t2 (t1 > t2 and t2 is not long enough for MTJ1 to switch), whereas in current-dependent writing, both MTJs are firstly written with larger write current (I_{write1}). Subsequently, the faster MTJ2 is programmed with smaller write current (I_{write2}) ($I_{write1} > I_{write2}$ and I_{write2} is not large enough for MTJ1 to switch).

Releasing the two MTJs with different I_c can be achieved by having different MTJ free layer thickness, dimensions [17], and/or width of underlying electrodes [15]. Therefore, P-MLC manufacturing might be challenging due to this imposed nonuniformity in cell architecture. The imposed different current requirements may result in lower energy efficiency (i.e. additional energy penalty) as one of the MTJs should switch with larger current amplitude compared to the other MTJ (under equivalent switching time assumption), whereas in conventional SLC SOT-MRAM, all the SOT-MTJs consume the same energy (i.e. no enforced rule of using two SOT-MTJs with different I_c). However, this energy penalty can be reduced by having a smaller ΔI_c between the two SOT-MTJs, as depicted in **Figure 3**. The minimum ΔI_c (minimum energy penalty) depends on the tolerable switching probability (PSW) of MTJ2 (slower SOT-MTJ) while writing MTJ1, as the smaller the ΔI_c, the higher the PSW of MTJ2. It is important to note that in real MRAM chip, the ΔI_c should be large enough to accommodate for the different distributions of switching time, switching current, and error rates. This should be chosen carefully by considering the potential variability of critical current ($\sigma I_c/\mu I_c$), which can be as large as 10% [13].

The reading operation is done similarly to the conventional SOT-MRAM, where the sense current or voltage is used to identify the equivalent resistance state. However, it is required to differentiate between four different resistances states unlike the only two states that exist in the conventional SLC. In addition, the

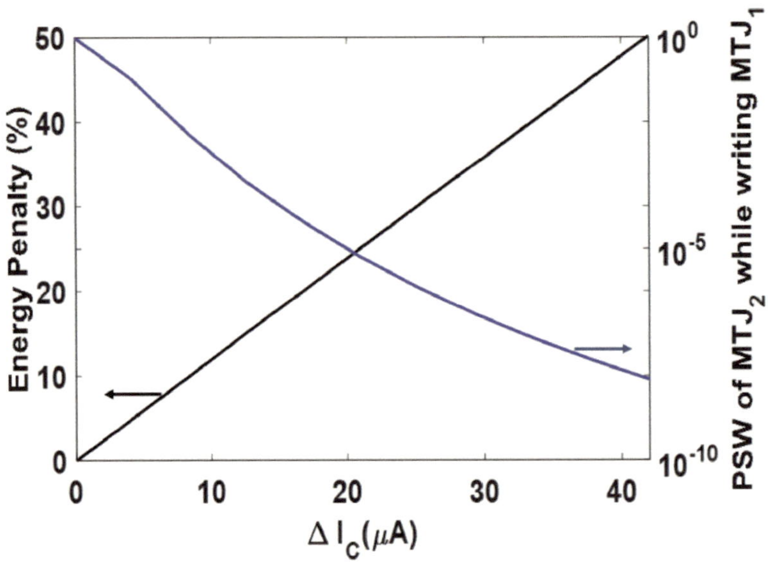

Figure 3.
Percentage of increase in the energy resulting from having two MTJs of different IC versus $\Delta I_c = I_{c2} - I_{c1}$ and the corresponding probability of switching (PSW) of slower MTJ2 while writing faster MTJ1 [17].

in-parallel configuration of the two MTJs during reading results in a reduced minimum difference between the various resistance states (and consequently reduced read margin compared to in-series configuration and SLC), as the equivalent resistance for two parallel resistances is always smaller than the smallest resistance. For instance, if the parallel configuration resistance (R_p) of MTJ1 (R_{p1}) =7 kΩ, R_p of MTJ2 (R_{p2}) = 12 kΩ, anti-parallel configuration resistance (R_{ap}) of MTJ2 (R_{ap2}) = 22 kΩ, the equivalent resistance for in-parallel MTJs connecting for case1 (R_{p1}, R_{p2}) is R_{tot1} = 4.4 kΩ and for case2 (R_{p1}, R_{ap2}) is R_{tot2} = 5.3 kΩ. That results in a minimum resistance difference (ΔR_{min}) of 0.9 kΩ only, while for SLC or even in-series MTJ connection, a larger ΔR_{min} can be achieved. For instance, if the MTJ connected in-series instead of in parallel using the same MTJ resistance values, the R_{tot1} becomes equal to 19 kΩ and R_{tot2} = 29 kΩ, which results in ΔR_{min} of 10 kΩ. Consequently, the combination of both MLC scheme and in-parallel connectivity increases the expected BER.

2.2.2 Series multi-level cell SOT-MRAM (S-MLC SOT-MRAM)

S-MLC, shown in **Figure 2(d)**, consists of two in-series MTJs placed over an electrode made of heavy metal. The first MTJ (MTJ1) in contact with the heavy metal electrode can be programmed by SOT effect, whilst the second MTJ (MTJ2) (stacked over MTJ1) must be programmed by conventional spin-transfer torque (STT). Similarly, the main advantage of this proposal is that each cell comprises two bits that are accessed by two transistors. This results in a nearly 50% reduction of the 1-bit effective area compared to conventional SOT-MRAM. In the write operation of S-MLC, MTJ2 must be programmed before MTJ1 to avoid a final state of write disturb failures for MTJ1, which means that the programming for the two MTJs should be serial and cannot be simultaneous. The need for STT in programming MTJ2 results in low energy efficiency as STT programming requires passing current by the high resistance MTJ stack, which demands high writing voltage in addition to the large critical current of STT switching compared to SOT switching. In addition, passing a large current through the MTJ stack reduces the tunnel barrier reliability, which jeopardizes one of the main advantages of using SOT-MRAM. The reading operation of the S-MLC is similar to P-MLC. Being an MLC requires the stack to represent four different resistance states. Thus, S-MLC uses MTJs with different cross-sectional dimensions to achieve the four different resistance states, which may result in a complex fabrication process. Furthermore, there is a need to employ low resistance MTJs in the stack to be able to supply enough current to achieve the STT switching. This results in a smaller minimum difference between the four distinct resistance states (i.e. smaller ΔR_{min} between the four possible resistances (R11, R10, R01 and R00)). The reduced ΔR_{min} minimizes the read margin for the S-MLC memory and thus longer reading delay.

3. Diode-based SOT-MRAM

This section presents the various diode-based or selector-based SOT-MRAM proposals in the literature. These proposals mainly rely on replacing the read transistor (Tx), highlighted in **Figure 2(a)**, by a diode or selector. As the read operation in SOT-MRAM requires mainly a relatively small and unidirectional current, a diode or a selector can successfully satisfy these requirements. The employed diodes are targeted to be nonsilicon-based, and thus, the cell silicon area would have one less transistor, as further explained below. However, employing a diode or selector may come with an energy penalty as larger read voltage may be required to overcome the diode's on-voltage.

3.1 SLC diode-based SOT-MRAM

Seo et al. [14] proposed a diode-based single-level cell (SLC) SOT-MRAM, shown in **Figure 2(b)**. In that design, the SOT-MRAM cell area is reduced by replacing the read transistor, depicted in **Figure 2(a)**, with a Schottky diode. Thus, the cell requires only one transistor to access a single bit. This design's main advantage is the reduction of the 1-bit effective area by approximately 50% compared to conventional SLC SOT-MRAM that requires two transistors to access a single bit. In this design, the 1-bit effective area is estimated to be $34.5F^2$ using the design rules in [16]. The write and read operations in this proposal are similar to that of the conventional SLC SOT-MRAM. The write operation would consume similar write energy, as each cell comprises only 1-bit (MTJ). Moreover, the read operation follows the same biasing as in the conventional SLC SOT-MRAM, where the RWL of the required row is activated and the WWL is deactivated. However, the diode usage increases the read energy as the read voltage needs to account for the additional voltage drop across the diode. In addition, there is still a margin to achieve smaller 1-bit effective area by using the diode as it is shown in the following sections.

3.2 Multi-bit per cell dedicated diode (MBC-DD) SOT-MRAM

Ali et al. [18] extended the SLC diode-based SOT-MRAM proposal to be a multi-bit per cell (MBC). Their proposed design, shown in **Figure 4**, relies on a metal-oxide-metal (MOM) diode (or also known as selector) stacked over an MTJ (forming a D-MTJ) to replace the task of the read Tx in controlling the read current flow. The cell comprises two D-MTJs with similar tunnel oxide barrier thicknesses and their free-layers are placed in contact with a common HM electrode. The two D-MTJs can be programmed with two different bits through a common electrode. The sharing of common electrode allows the use of only single transistors to access

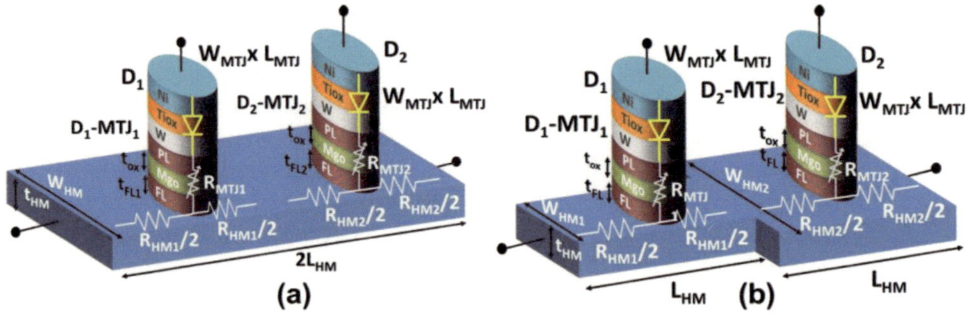

Figure 4.
3D structure of the proposed dedicated diode multi-bit per cell (MBC) DD SOT-MRAM with (a) uniform HM electrode and MTJs with different t_{fl} *[18], (b) different HM width and MTJs with similar* t_{fl} *[18].*

the two bits (2 D-MTJs). This result is the main advantage of this proposal, which is a reduced 1-bit effective area that is 4× smaller than the conventional SOT-MRAM and offers at least double the density compared to any MRAM proposal in the literature. Given the design rules in [16], the 1-bit effective area is estimated to be $18F^2$. The cell also employs separate read word line (RWL) for each D-MTJ, as in **Figure 5**. On the other hand, similar to [14], employing a dedicated diode per MTJ may increase the read energy and limit the maximum acceptable diode area.

The MOM diode MTJ stack employed in this design is similar to the experimentally validated device used in the 1S1R 3D cross-point STT-MRAM developed by avalanche [19–21]. However, employing the D-MTJ device in the SOT technology would be more efficient than in the STT technology as the energy and performance degradation effects of the diode would only exist in the read operation and is avoided in the write operation. This is because the diode only exists in the read path, while in the STT technology, the write operation is dependent on the employed diode as the write current has to flow through both the MTJ and diode to achieve the STT switching.

The writing operation of this cell is similar to that in the P-MLC, where either time-dependent or current-dependent writing can be adopted as elaborated before. During the write operation, the WWL of the row comprising the required cell is asserted high. A '0'/'1' is written on the MTJ if the BL is set high/low, and SL is pulled low/high for the column including the targeted cell, allowing the charge current to flow through the HM electrode in the essential direction, as shown in **Figure 5(a)**. Furthermore, the RWLs of the row comprising the targeted cell are set low to ensure that the diodes are reverse-biased and no leakage current flows

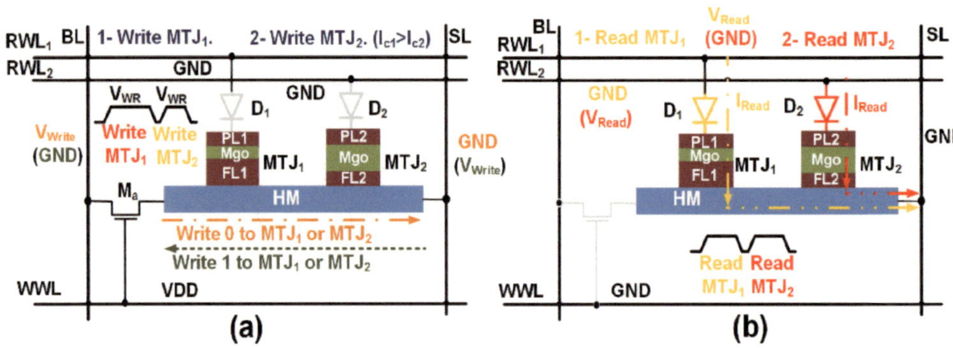

Figure 5.
Proposed MBC-DD SOT-MRAM cell in the (a) write and (b) read operations [18].

through the MTJs. It is worth mentioning that the low voltage while using the MOM diode may not be exactly zero volts, as a hold voltage ($V_{hold} \sim 0.02$ V) might be needed.

The separate RWLs for each of the two D-MTJs permit selective reading of the two bits. During the read operation, the WWL is deactivated, and the SL of the column comprising the targeted D-MTJs is pulled to GND. The RWL of the targeted D-MTJ is then connected to the sense amplifier to forward bias the diode and read out the data stored in the MTJ, as shown in **Figure 5(b)**. Although switches are still needed to select different RWL, it has a negligible impact on the area per unit cell as these switches are shared by all the cells in the row. Furthermore, as both MTJs are sensed independently and the RMTJ is an order of magnitude larger than RHM, hence, the RHM impact on the effective TMR is minimized (average sensed resistance = RHM + RMTJ).

The realization of the MBC-DD SOT-MRAM has two main essential requirements. Firstly, employing two SOT-MTJs with different I_c to be able to write the two bits independently as discussed before. This is achieved by using either different widths of the HM below each MTJ (W_{HM}), as illustrated in **Figure 4(b)**, or using different free layer (FL) thicknesses (t_{fl}) within each MTJ, as shown in **Figure 4(a)**. Uniform HM would be preferred in high-density memories, as W_{HM} for both SOT-MTJs are limited to the technology minimum feature size (F), and any increase in the W_{HM} than F increases the overall cell area. However, it may require two MTJ stack deposition with additional lithography steps, which may increase the manufacturing cost, whereas SOT-MTJs with uniform t_{fl} is preferred from reading perspective as both MTJs would have similar TMR [22], which simplifies the read operation. It also offers lower fabrication cost due to simple processing. Nevertheless, due to the higher density, at least double other designs, it is expected that both structures would lead to lower overall cost per bit cell. Secondly, a 3D diode MTJ stack is required to have the diode replacing the read transistor without consuming silicon area. Incorporating the diode in the reading process requires applying a relatively higher read voltage (V_{Read}) that is enough to overcome the diode on-voltage (i.e. forward bias the diode) and supply the required read current. The required magnitude of the V_{Read} at a targeted read current (I_{Read}) depends mainly on two factors, which are the diode's cross-sectional area and the load resistance (R_{load}). Firstly, the smaller the diode's area, the smaller the diode supplied current at given bias voltage, as shown in **Figure 6(a)**. Thus, the needed V_{Read} increases with diode scaling down to supply the targeted I_{Read}, as depicted in **Figure 6(b)**. Secondly, the load resistance (R_{load}) affecting the diode, which is the in-parallel SOT-MTJs equivalent resistances and the existing transistors in the I_{Read} path. The smaller the R_{load}, the smaller the required V_{Read} to supply certain I_{Read} for a given diode area, as illustrated in **Figure 6(c)** and modeled in Eq. (1). Thus, smaller R_{load} aids to achieve lower read energy at a given I_{Read}.

$$V_{Read} = V_{On\_noload} + I_{Read} * R_{load}, \tag{1}$$

where V_{on_noload} is the diode's required bias voltage to supply certain I_{Read} with no load condition.

3.3 Multi-level cell shared diode (MLC-SD) SOT-MRAM

Ali et al. [17] also proposed another diode-based SOT-MRAM cell in which a shared diode (selector) between the two MTJs in the cell is employed, as illustrated in **Figure 7**, instead of a dedicated diode (selector) per each MTJ. Similar to MBD-DD SOT-MRAM, each cell comprises 2-bits (MTJs) accessed with a single transistor.

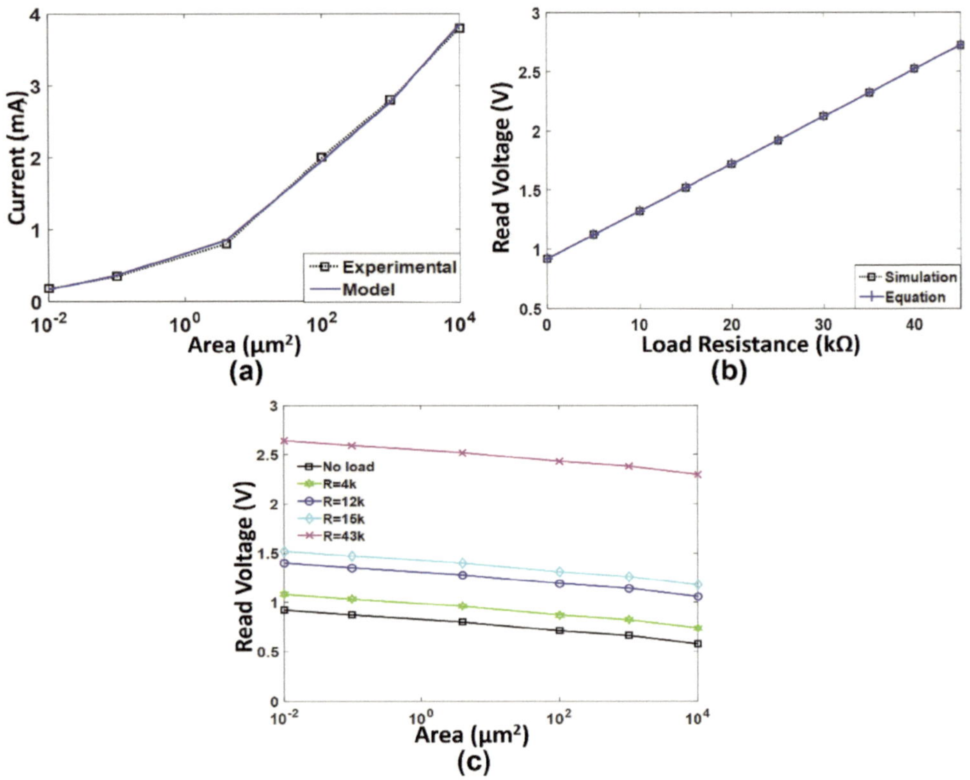

Figure 6.
(a) The diode current versus diode area @ bias voltage = 1 V and no-load resistance. The data are from the implemented diode Verilog-A model verified against experimental data in [23]. (b) The required read voltage to supply I_{Read} *= 40 μA with diode area scaling for various* R_{load}*. (c) The required read voltage for the diode to supply* I_{Read} *= 40 μA versus different* R_{load} *using both Verilog-A model simulation and Eq. (1) [18].*

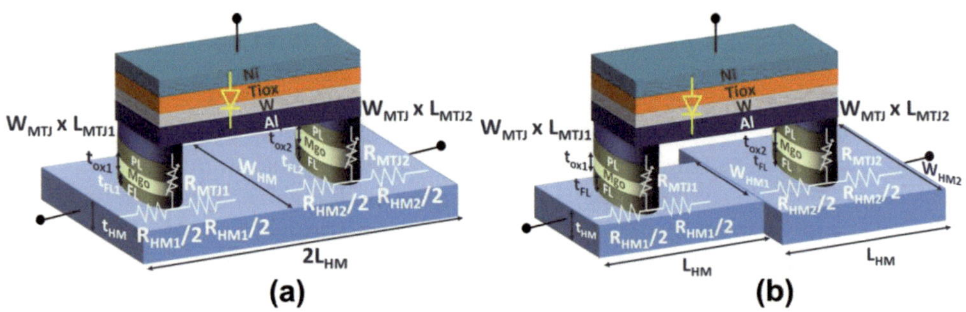

Figure 7.
3D structure of the proposed shared diode multi-level cell (MLC) SOT-MRAM with (a) uniform HM electrode and MTJs with different t_{fl} *[17], (b) different HM width and MTJs with similar* t_{fl} *[17].*

This result is a similar 1-bit effective area that is 4× smaller than conventional SOT-MRAM and at least double the density compared to any of the MRAM design in the literature. The 1-bit effective area is estimated to be $17.5F^2$ using the design rules in [16]. On the other hand, employing a shared diode permits increasing the diode area as it is no longer limited by one MTJ area. However, the needed memory sensing will be MLC (i.e. the cell has four different resistance states) instead of a SLC (i.e. the cell has only two resistance states), which complicates the sensing operation and reduces the read margin.

The two MTJs in the MLC-SD cell also share a common HM electrode to be able to program the two MTJs with the energy-efficient SOT technology. This enforces using two SOT-MTJs with different I_c as well, which can be written following the same approach of P-MLC using either time or current-dependent writing. Consequently, similar to MBC-DD in the write operation, the WWL of the row comprising the required cell is asserted high. A '0'/'1' is written to the MTJ if the BL is set to high/low and SL is pulled to low/high for the column including the targeted cell, allowing the charge current to flow through the HM electrode in the essential direction, as depicted in **Figure 8(a)**. In addition, the RWL of the row comprising the targeted cell are set low to ensure that the diodes are reverse biased and no leakage current flows through the non-selected cells.

Employing a shared diode requires connecting the two MTJs in-parallel, which enforces the cell to be MLC. Consequently, the two MTJs in the cell should have different R_P and R_{AP} such that their equivalent resistance has four distinct values, which represents two-bit logic values of '00', '01', '10' and '11'. To read the bits in the cell, the WWL and the SL signals are pulled low to deactivate all the write transistors and allow the read current to flow through the targeted MTJs. The row RWL is connected to the sense amplifier and consequently is pulled up to the required read voltage (V_{Read}), as shown in **Figure 8(b)**. Thereafter, the equivalent 2 bits for the targeted MLC cell are identified sequentially by comparing the currents flowing through the sensed combined MTJs and the appropriate reference MTJs using a binary search algorithm [7]. In this algorithm, the sensed MTJs are first compared with the first reference resistance to determine the first-bit state. Based on the first-bit state, one of the two other reference resistances are then chosen to compare with the sensed MTJs to determine the second-bit state. The same technique can be used to sense the state of any of the MLC proposals illustrated above.

As can be inferred from above, the realization of MLC-SD SOT-MRAM has three main essential requirements. Similar to the MBC-DD SOT-MRAM design, it requires employing two SOT-MTJs with different I_c and a 3D diode MTJ stack. In the MLC-SD cell, as the diode is shared, the employed diode can be large, and its size can be approximately up to the whole cell area instead of just one MTJ area. A larger diode requires a smaller read voltage to supply the required I_{Read}, as depicted in **Figure 6(a)**. Moreover, the in-parallel combination of the two MTJs results in smaller overall resistance compared to a single MTJ resistance. This reduces the

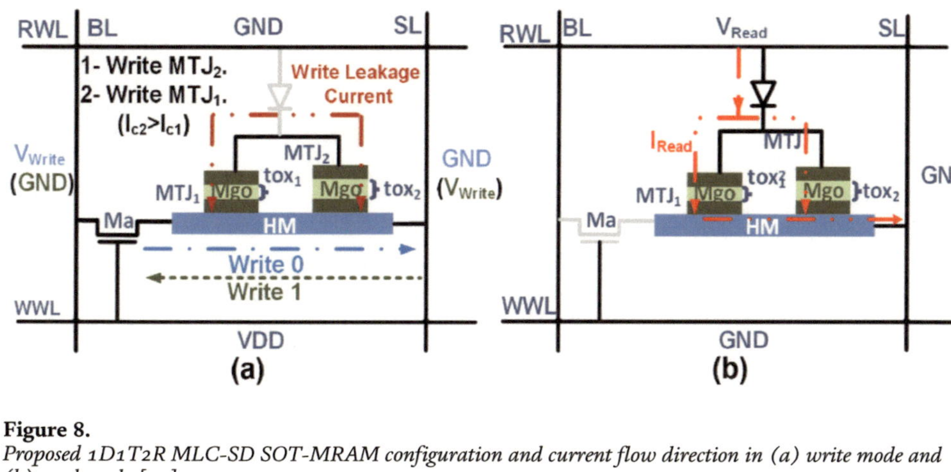

Figure 8.
Proposed 1D1T2R MLC-SD SOT-MRAM configuration and current flow direction in (a) write mode and (b) read mode [17].

diode's R_{load}, which further decreases the required V_{Read}, as shown in **Figure 6(b)**. Hence, the two factors that affect the required V_{Read} (i.e. diode area and R_{load}) are improved in this design compared to the MBC-DD and SLC diode-based SOT-MRAM designs. Hence, the smaller required V_{Read} may permit the MLC-SD design to achieve smaller read energy consumption. Furthermore, MLC-SD cell requires employing two MTJs with different R_P and R_{AP} values as it is an MLC. Assuming that both the MTJs use the same materials, different MTJ resistances are achieved by varying either the MTJ dimensions (i.e. W_{MTJ} and L_{MTJ}), the dielectric thickness (t_{ox}), or combination of both [24]. It is essential to have a large minimum resistance difference (ΔR_{min}) between the distinct in-parallel equivalent resistance states to increase the distinguishability and read speed. However, the MLC would mainly have smaller ΔR_{min} compared to the SLC, which in-return may result in reduced reading speed and would be a competing factor with the reduced V_{Read} to decide the read energy efficiency of this cell compared to the SLC proposals.

4. Evaluation

In this section, the various proposals in the literature are evaluated using the same SOT-MTJ technology in [13] on both cell and system level perspectives. The cell level analyses are done based on a 2 × 2 memory array. The simulations run over Cadence Virtuoso and using the SOT-MTJ Verilog-A model demonstrated in [25] and the parameters in **Table 1**. The system-level analyses are done using the non-volatile memory simulator, known as NVSim [26]. NVSim estimates the overall memory performance, power consumption, energy, and area based on the given memory cell parameters.

Symbol	Parameter	Value
$W_{MTJ} \times L_{MTJ}$	MTJ dimensions ($W \times L$) (nm^2)	50 × 100
t_{FL}	Free layer thickness (nm)	1.5
K_uV/KT	Thermal stability	46
t_{ox}	Tunnel barrier thickness (nm)	1.8
M_s	Magnetization saturation (emu/cm^3)	1114
R_{AP}	MTJ high resistance value (kΩ)	15
TMR	Tunnel magneto-resistance ratio (%)	114
α	Gilbert damping	0.012
ρ_{HM}	Heavy metal (W) resistivity (μΩ·cm)	200
$W_{HM} \times L_{HM}$	HM dimensions ($W \times L$) (nm^2)	115 × 150
t_{HM}	Heavy metal thickness (nm)	3
J_{co}	Critical current density ($\times 10^{10}$ A/m^2)	7
θ_{SHE}	Spin hall angle	0.3
P	Spin polarization	0.6
γ	Gyromagnetic ratio (rad s^{-1} T^{-1})	1.76×10^{11}
$\hbar$	Reduced Planck constant (J s)	1.054×10^{34}

The parameters follow the experimental data of the SOT-MTJ in [13] and are explained in [11].

Table 1.
SOT-MTJ device parameters.

4.1 Cell-level evaluation

Table 2 presents a comparison between the different MRAM designs. In terms of area, MBC-DD and MLC-SD SOT-MRAM do offer the smallest 2-bit cell area among the various designs, which are estimated to be $36F^2$ and $34.5F^2$, respectively, based on the rules in [16]. This is at least double the density compared to other MRAMs and achieves 75% smaller 1-bit effective area compared to conventional SOT-MRAM. From energy perspective, these designs consume at least 36% less energy compared with designs utilizing STT writing (S-MLC), due to the high energy efficiency of SOT writing. Unlike P-MLC and MLC-SD SOT-MRAM, MBC-DD has no write current leakage through the MTJ from the HM during write mode as the diodes are reverse-biased, which also leads to better energy efficiency. However, the significant area reduction for both MBC-DD and MLC-SD SOT-MRAMs comes with additional energy penalty in both worst-case write operation (i.e. writing non-identical bits) and read operation. The additional energy consumption in the worst-case write operation of non-identical bits is because of the enforced rule of using two MTJs per cell with different I_c. For instance, writing two different bits ('10' or '01') in MBC-DD and MLC-SD consume higher energy (0.88 pJ with 10.5 ns delay) compared to writing two SLC SOT-MRAM (0.76 pJ). However, to write identical bits ('00' or '11') on the MBC-DD and MLC-SD cells require only a single write pulse, which leads to better energy efficiency than SLC SOT-MRAM. This is because programming two identical bits in the SLC SOT-MRAM always requires two write pulses. Thus, if equal probability of programming '00', '01', '10', and '11' is assumed, MBC-DD and MLC-SD designs may result in similar average energy efficiency to two bits of SLC SOT-MRAM, as shown in **Table 2**. Moreover, the

		SLC SOT (2-bit)	1D1T SLC-SOT (2-bit) [14]	S-MLC [15]	P-MLC [15]	MLC-SD SOT-MRAM [17]	MBC-DD SOT-MRAM [18]
Energy per 2-bit (pJ)	Write (EW)[a] (w. case)	0.76	0.76	1.13	0.72 (0.88)	0.72 (0.88)	0.72 (0.88)
	Write leakage	0.0002	0.0001	0.0001	0.19	0.2	0.0001
	Read (Er)	0.024	0.036	0.034	0.039	0.041	0.036
Total energy/2-bit		0.78	0.8	1.17	1.11	1.11	0.9
Delay per 2-bit (ns)	Write a (w. case)	9	9	10.5	8.1 (10.5)	8.1 (10.5)	8.1 (10.5)
	Read	1	1	1.4	1.6	1.6	1
Read voltage (V)		1.2	1.8	1.2	1.2	1.3	1.8
ΔR_{min} (kΩ)[b]		5	5	3	1.4	1.4	5
Diode area (F^2)[d]		—	24	—	—	24	10
Area (A)/1bit (F^2)[c]		69	34.5	50	34.5	17.25	18
FOM (Er*Ew*A)		2.19×	1.65×	5.7×	2.5×	1.34×	1

[a]*Energy and delay are the average of writing the data '00', '01', '10', '11'.*
[b]ΔR_{min} *is* $R_{AP} - R_P$ *for SLC, while it is the minimum resistance difference among the four different states in MLC.*
[c]*Area estimated based on standard design rules reported in [16].*
[d]F *is the minimum feature size of the employed technology (i.e. In 32 nm technology,* F *= 32 nm).*

Table 2.
Comparison of various MRAM technologies.

energy penalty for non-identical bits writing can be also minimized as discussed before. In terms of reading, MBC-DD maintains similar distinguishability to conventional SLC, as indicated by the ΔR_{min} values in **Table 2**, because each of the two MTJs is sensed separately. This is unlike the P-MLC, S-MLC, and MLC-SD structures that have a reduced ΔR_{min} as a result of relying on an MLC approach. However, to ensure sufficient diode drive current within the D-MTJs, a larger read voltage is needed compared to cells with read transistors, which does increase the read energy consumption compared to conventional SLC SOT-MRAM.

The diode-based SLC SOT-MRAM design [14] does offer the advantage of maintaining a similar write energy efficiency compared to the conventional SOT-MRAM (i.e. baseline from write energy perspective). Moreover, it offers 50% 1-bit effective area savings compared to conventional SOT-MRAM, which is on level with P-MLC and S-MLC designs with the advantage of maintaining an SLC sensing approach. However, diode-based SLC SOT-MRAM still consumes double the area compared to the MBC-DD and MLC-SD designs, while it still also suffers from the energy and fabrication complexity penalty of employing a diode.

Similar to diode-based SLC SOT-MRAM, P-MLC design offers 50% 1-bit effective area savings compared to conventional SOT-MRAM, whereas S-MLC does offer only 28% savings as the transistor size needs to increase to supply the required STT current through the high resistance MTJ stack. On the other hand, both P-MLC and S-MLC do not employ a diode, which may aid in reducing the read energy and avoiding the fabrication process issues related to incorporating a diode. However, they suffer from other drawbacks that increase both energy consumption and fabrication complexity. In particular from energy perspective, S-MLC consumes high energy due to using STT technology in writing one of the two MTJs per cell, whereas P-MLC shares the additional write energy penalty issue in writing non-identical bits per cell with MBC-DD and MLC-SD designs. In addition, both of the designs rely on the MLC sensing approach, which harms the sensing speed and distinguishability as reflected by the reduced ΔR_{min} values.

Overall, if a figure-of-merit (FOM) is defined as the product of the area, read and write energy products [27], MBC-DD SOT-MRAM may outperform other designs by at least 34%, thanks to its significant area reduction and maintaining the SLC sensing approach.

4.2 System-level evaluation

As aforementioned, NVSim [26] is used to evaluate the various SOT-MRAM cells from a system-level perspective. NVSim does consider the different write/read peripherals, array organization, and routing network required in the overall memory architecture. NVSim supports various nonvolatile memories such as STT-MRAM, PCRAM, and ReRam in addition to the well-known volatile memories such as SRAM and DRAM. NVSim can be tuned to support SOT-MRAM as well. In this study, the comparison is based on the utilization of the various nonvolatile MRAM cells as cache memory and they are mainly compared to the current widely used technology as cache memory, which is the SRAM. SRAM does offer high performance; however, currently, it suffers from a significant increase in the 1-bit area and the leakage power consumption [28]. Thus, the utilization of an area-efficient, high-speed and nonvolatile SOT-MRAM would be a promising solution to replace the existing SRAM technology, especially in higher-level caches. The considered cache has 4-way set associativity and 64 Byte line size and is optimized to achieve the smallest overall silicon area. The 32 nm technology node is assumed for the various designs, in which the SRAM cell area is $170F^2$ [28], and the same cell parameters for the various MRAM technologies are maintained as stated above.

To clarify the pros and cons of the various memory technologies, three different memory capacities are considered, which are 256 KB, 1 MB, and 8 MB. **Figure 9** depicts the total leakage power consumption for the various designs with three different capacities. The volatile SRAM technology does consume significant leakage power, which increases by order of magnitudes for larger memory capacities. On the other hand, the nonvolatility of the MRAM designs results in relatively negligible leakage among the various capacities. This demonstrates the advantage of significant power consumption reduction by employing the nonvolatile MRAM designs as a replacement of the volatile SRAM, especially in battery-powered mobile devices that demand long idle durations.

From area perspective, **Figure 10** shows the comparison between the various memory technologies, estimated by NVSim, while being used as cache memory. **Figure 10(a)** reports the different MRAM designs relative to the overall SRAM memory area for the three different capacities. The figure indicates the significant reduction in the overall area by employing the various MRAM technologies in comparison to SRAM, which consumes at least 50% smaller silicon area. Furthermore, the area saving percentage increases noticeably for larger memory capacities. This is because for larger memory capacity, the impact of the cell area overtakes the impact of the periphery. The periphery area for MRAM consumes larger area than the periphery area of the SRAM due to the need for relatively larger write currents, whereas the cell area of the MRAM is much smaller than the SRAM cell area [29]. Hence, at larger capacities, the MRAMs that have significantly smaller cell area compared to SRAM would consume much lower overall silicon area. For instance, the area saving for the MBC-DD and MLC-SD cells, which are the cells with smallest footprints, can reach up to 90% smaller area compared to SRAM for large cache memory size, such as 8 MB size.

The impact of the smaller cell area is also clear while comparing the various MRAM technologies, as presented in **Figure 10(b)**. The smaller 1-bit effective area of S-MLC, SLC diode-based, P-MLC, MLC-SD, and MBC-DD SOT-MRAMs by 28, 50, 50, 75, and 74% compared to the conventional SOT-MRAM results in similar overall memory silicon area reduction. In particular, for larger memory capacity (e.g. 8 MB), where the impact of the cell area is more significant, the overall memory silicon area of the various designs in the literature relative to conventional SOT-MRAM design shows approximately equal area reduction as the 1-bit effective area reduction. Moreover, the smaller silicon area consumption of the various proposed designs in the literature compared to SRAM and conventional SOT-MRAM with equivalent capacity permits realizing the cache memory using these cells with higher capacity under the iso-area assumption. For instance, 2 MB

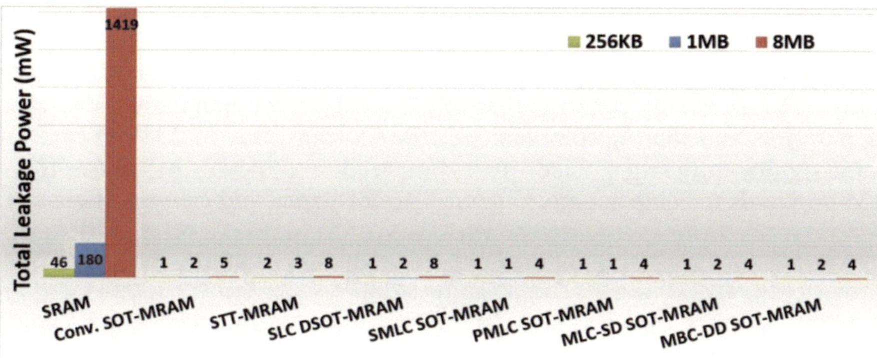

Figure 9.
Total leakage power consumption of SRAM, STT-MRAM, conventional, SLC DSOT-MRAM, S-MLC, P-MLC, MLC-SD, and MBC-DD SOT-MRAMs for various memory capacities.

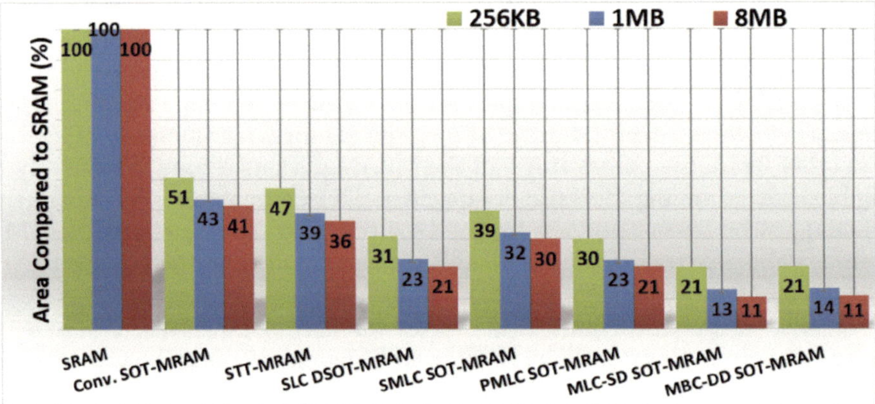

(a)

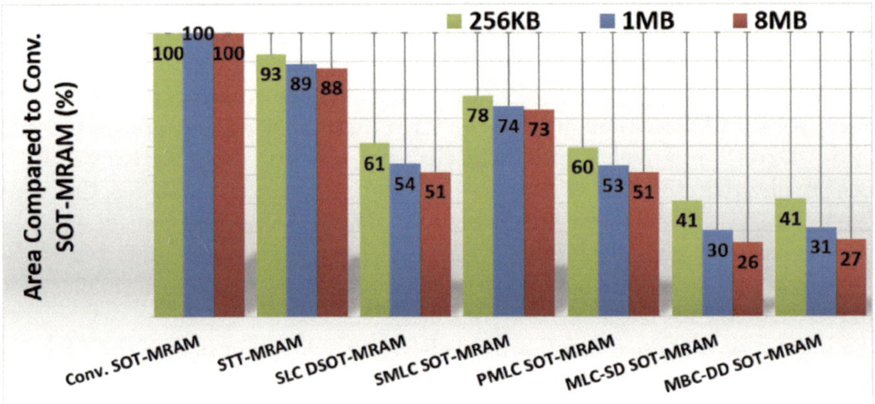

(b)

Figure 10.
Area comparison of the STT-MRAM, conventional, SLC diode-based, S-MLC, P-MLC, MLC-SD, and MBC-DD SOT-MRAMs for different capacity relative to (a) SRAM and (b) conventional SOT-MRAM.

capacity of conventional SOT-MRAM consumes similar silicon area to an 8 MB capacity of MBC-DD SOT-MRAM. Higher memory capacity results in higher performance metrics, such as instruction per cycle (IPC) and energy efficiency due to reduced access counts for the off-chip memory [30].

From energy and performance perspective, SRAM can have higher hit/miss performance and energy efficiency compared to the various SOT-MRAM proposals at small memory capacity (e.g. 256 KB), as depicted by **Figure 11(a)**. This is because the impact of the large load capacitance of the SRAM cell would be minimal at these capacities [29]. However, for large memory capacity, the much higher load capacitance, parasitic, and routing complexity of the six transistors SRAM cell (as SRAM consume significantly larger silicon area compared to SOT-MRAM) causes the MRAM proposals to achieve better hit/miss performance and energy efficiency. On the contrary, the write energy for the various MRAM proposals is larger than that of the SRAM for the various memories capacity, as shown in **Figure 11(b)**. This is attributed to the larger write current requirement for the MRAM-based technologies relative to the SRAM technology. However, with improved SOT-MRAM technology such as the type-x SOT-MTJ [13] (achieves sub ns switching with write current of 100 μA), or the presented high-performance SOT-MRAM by IMEC [12] that achieves successful switching in 100's of ps range, the write energy can be reduced significantly, as illustrated in **Figure 12**. This makes the various

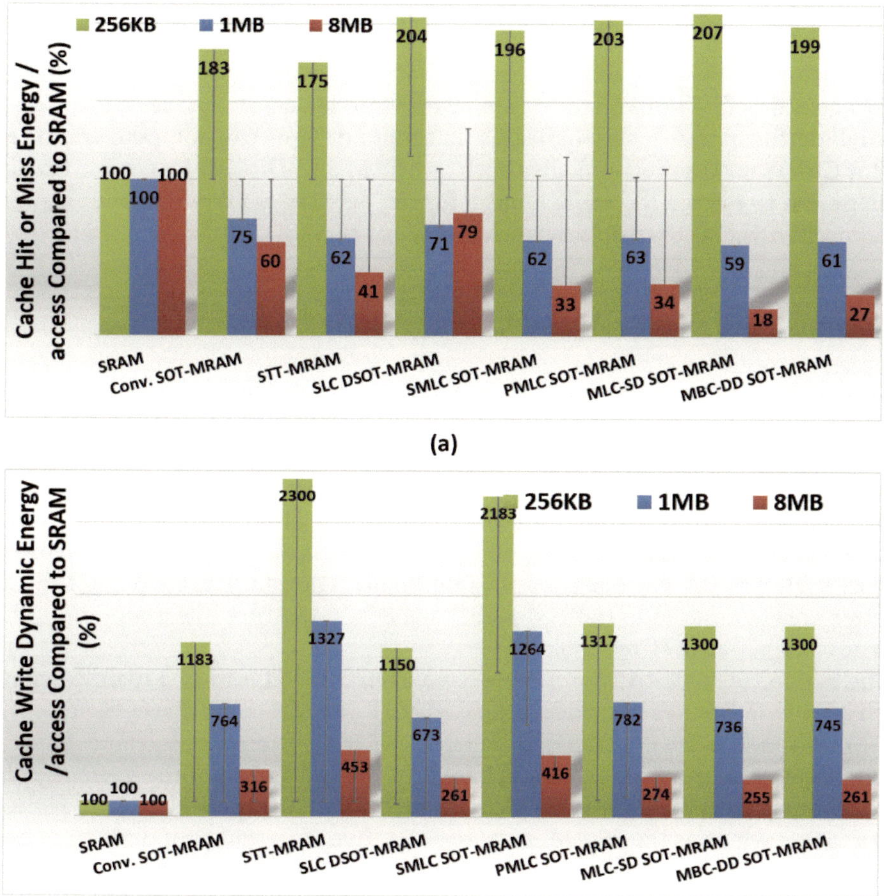

(a)

(b)

Figure 11.
Comparison of the SRAM, STT-MRAM, conventional, SLC diode-based, S-MLC, P-MLC, MLC-SD, and MBC-DD SOT-MRAMs for different capacities relative to SRAM from (a) cache hit/miss energy per access perspective and (b) cache write dynamic energy per access perspective.

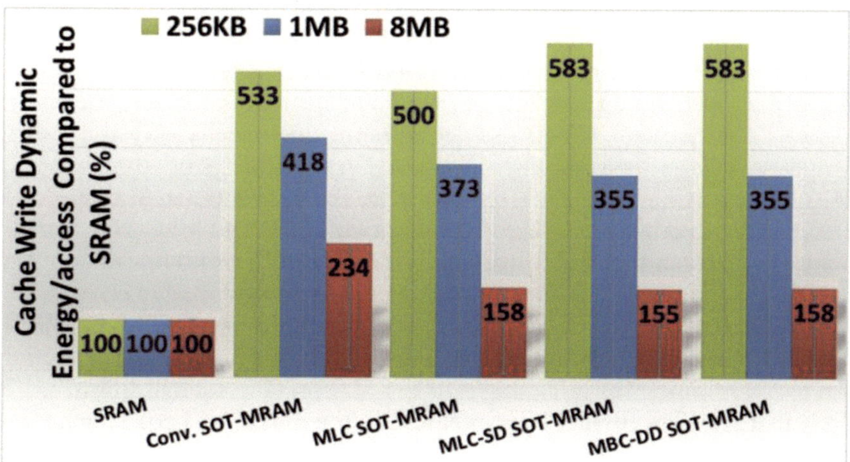

Figure 12.
Cache write dynamic energy per access comparison of the SRAM, conventional, P-MLC, MLC-SD, and MBC-DD SOT-MRAMs for different capacity relative to SRAM assuming improved high-speed SOT-MTJ technology (e.g. type-x reported in [13]).

SOT-MRAM proposals a viable and realistic solution to replace the SRAM technology in certain applications.

In conclusion, the previous discussion shows that the various proposed SOT-MRAM cells do offer nonvolatility (i.e. nearly zero leakage), smaller silicon area, and high performance. It also indicates that these designs can compete with the current CMOS volatile technologies such as SRAM and DRAM. However, further development to reduce the write energy for the existing SOT-MTJ technology may be required to widen the application window for such SOT-MRAM technologies.

5. Conclusion

This chapter presents the various SOT-MRAM proposals in the literature highlighting the pros, cons, and operation of each design. SOT-MRAM relies on SOT technology, which offers various advantages such as high energy efficiency, fast switching speed, and high device reliability. However, conventional SLC SOT-MRAM requires two transistors to access a single bit. This in return results in a relatively large 1-bit effective area, which limits its application for large memory capacities. Hence, the various proposals in the literature targets reducing the 1-bit effective area compared to both conventional SOT-MRAM while maintaining the main advantages of SOT technology.

The various SOT-MRAM proposals have been divided into two main categories, which are diode-based and nondiode-based SOT-MRAM cells. These various SOT-MRAM cells have been evaluated from both cell and system level perspectives. The system-level evaluation is performed based on the utilization of the various cells as cache memory, and they are mainly compared to the current widely used technology as cache, which is the SRAM. In particular, five different proposals have been investigated. These proposals are S-MLC, P-MLC, diode-based SLC, MBC-DD, and MLC-SD SOT-MRAMs that are shown to offer 70, 79, 79, 89, and 89% reduced 1-bit effective area compared to SRAM and 28, 50, 50, 74, and 75% compared to conventional SOT-MRAM, respectively.

From energy perspective, S-MLC, P-MLC, MBC-DD, and MLC-SD consume higher write energy compared to conventional SOT-MRAM. P-MLC, MBC-DD, and MLC-SD consume higher worst-case write energy while writing nonidentical bits on the cell due to the enforced rule of employing two SOT-MTJs with different I_C. However, if an equal probability of programming the various bits options is assumed, average energy efficiency similar to conventional SOT-MRAM may be achieved, whereas S-MLC involves writing one of the two MTJs in the cell using the energy-inefficient STT technology, which also degrades the device reliability as a result of supplying large current through the MTJ stack. On the other hand, diode-based SLC SOT-MRAM may achieve similar write energy to the conventional SLC SOT-MRAM. However, similar to other diode-based designs (MBC-DD and MLC-SD), it still incorporates a diode in the read operation, which may add additional energy penalty as the read voltage needs to be large enough to overcome the diode's on-voltage.

It is noteworthy that the SLC proposals such as diode-based SLC and MBC-DD would be preferred solutions, thanks to their offered significant area reduction in addition to maintaining the advantages of the SLC sensing such as improved BER. However, that requires further improvement in the diode-MTJ stack technology such that the diode transient response would match the required read performance and the diode's on voltage would be small, and thus, the read energy will be reasonable. On the other hand, the MLC proposals such as P-MLC, S-MLC, and MLC-SD require improvements in the MLC sensing techniques to enhance its sensing distinguishability and BER such that it will meet the industry standards.

Author details

Karim Ali
National University of Singapore, Singapore

*Address all correspondence to: karim.ali@u.nus.edu

References

[1] Alioto M. Enabling the Internet of Things: From Integrated Circuits to Integrated Systems. Cham: Springer International Publishing; 2017

[2] Raha A, Jaiswal A, Sarwar SS, Jayakumar H, Raghunathan V, Roy K. Designing energy-efficient intermittently powered systems using spin-hall-effect-based nonvolatile SRAM. IEEE Transactions on Very Large Scale Integration (VLSI) Systems. 2018;**26**(2):294-307

[3] Sakimura N, Sugibayashi T, Nebashi R, Kasai N. Nonvolatile magnetic flip-flop for standby-power-free SoCs. IEEE Journal of Solid-State Circuits. 2009;**44**(8):2244-2250

[4] Ali K, Li F, Lua SYH, Heng CH. Compact spin transfer torque non-volatile flip flop design for power-gating architecture. In: 2016 IEEE Asia Pacific Conference on Circuits and Systems (APCCAS). 2016. pp. 119-122

[5] Chabi D et al. Ultra low power magnetic flip-flop based on checkpointing/power gating and self-enable mechanisms. IEEE Transactions on Circuits and Systems I: Regular Papers. 2014;**61**(6):1755-1765

[6] Wang W-G, Li M, Hageman S, Chien CL. Electric-field-assisted switching in magnetic tunnel junctions. Nature Materials. 2011;**11**:64

[7] Jabeur K, Di Pendina G, Bernard-Granger F, Prenat G. Spin orbit torque non-volatile flip-flop for high speed and low energy applications. IEEE Electron Device Letters. 2014;**35**(3): 408-410

[8] Jabeur K, Di Pendina G, Prenat G. Ultra-energy-efficient CMOS/magnetic non-volatile flip-flop based on spin-orbit torque device. Electronics Letters. 2014;**50**(8):585-586

[9] Kwon KW, Choday SH, Kim Y, Fong XY, Park SP, Roy K. SHE-NVFF: Spin hall effect-based nonvolatile flip-flop for power gating architecture. IEEE Electron Device Letters. 2014;**35**(4): 488-490

[10] Bishnoi R, Oboril F, Tahoori MB. Non-volatile non-shadow flip-flop using spin orbit torque for efficient normally-off computing. In: 2016 21st Asia and South Pacific Design Automation Conference (ASP-DAC). 2016. pp. 769-774

[11] Wang ZH, Zhao WS, Deng EY, Zhang Y, Klein JO. Magnetic non-volatile flip-flop with spin-hall assistance. Physica Status Solidi Rapid Research Letters. 2015;**9**(6):375-378

[12] Imec demonstrates manufacturability of state-of-the-art spin-orbit torque MRAM devices on 300mm Si wafers, 2018

[13] Fukami S, Anekawa T, Zhang C, Ohno H. A spin–orbit torque switching scheme with collinear magnetic easy axis and current configuration. Nature Nanotechnology. 2016;**11**(7):621-625

[14] Seo Y, Kwon KW, Roy K. Area-efficient SOT-MRAM with a Schottky diode. IEEE Electron Device Letters. 2016;**37**(8):982-985

[15] Kim Y, Fong XY, Kwon KW, Chen MC, Roy K. Multilevel spin-orbit torque MRAMs. IEEE Transactions on Electron Devices. 2015;**62**(2):561-568

[16] Gupta SK, Park SP, Mojumder NN, Roy K. Layout-aware optimization of STT MRAMs. In: 2012 Design, Automation & Test in Europe Conference & Exhibition (DATE). 2012. pp. 1455-1458

[17] Ali K, Li F, Lua SYH, Heng C. Area efficient shared diode multi-level cell

SOT-MRAM. IEEE Transactions on Magnetics. 2018;**54**:1-5

[18] Ahmed KA, Li F, Lua SYH, Heng C. Area-efficient multibit-per-cell architecture for spin-orbit-torque magnetic random-access memory with dedicated diodes. IEEE Magnetics Letters. 2018;**9**:1-5

[19] Huai Y et al. High density 3D cross-point STT-MRAM. In: 2018 IEEE International Memory Workshop (IMW). 2018. pp. 1-4

[20] Yang H et al. Threshold switching selector and 1S1R integration development for 3D cross-point STT-MRAM. In: 2017 IEEE International Electron Devices Meeting (IEDM). 2017. pp. 38.1.1-38.1.4

[21] Yang H et al. 3D cross-point spin transfer torque magnetic random access memory. SPIN. 2017;**7**(3):1740011

[22] Lim WC, Park BG, Bae JY, Lee TD. Effects of free layer materials and thickness on TMR behaviour in magnetic tunnel junctions. Physica Status Solidi. 2004;**201**(8):1696-1699

[23] Li YT et al. Novel self-compliance bipolar 1D1R memory device for high-density RRAM application. In: 2013 5th IEEE International Memory Workshop. 2013. pp. 184-187

[24] Wang ZH, Zhao WS, Deng EY, Klein JO, Chappert C. Perpendicular-anisotropy magnetic tunnel junction switched by spin-hall-assisted spin-transfer torque. Journal of Physics D: Applied Physics. 2015;**48**(6):1-7

[25] Ali K, Li F, Lua SYH, Heng C. Energy- and area-efficient spin–orbit torque nonvolatile flip-flop for power gating architecture. IEEE Transactions on Very Large Scale Integration (VLSI) Systems. 2018;**26**(4):630-638

[26] Dong X, Xu C, Xie Y, Jouppi NP. NVSim: A circuit-level performance, energy, and area model for emerging nonvolatile memory. IEEE Transactions on Computer-Aided Design of Integrated Circuits and Systems. 2012; **31**(7):994-1007

[27] Lee A et al. RRAM-based 7T1R nonvolatile SRAM with 2x reduction in store energy and 94x reduction in restore energy for frequent-off instant-on applications. In: 2015 Symposium on VLSI Circuits (VLSI Circuits). 2015. pp. C76-C77

[28] Daly DC, Fujino LC, Smith KC. Through the looking glass—the 2018 edition: Trends in solid-state circuits from the 65th ISSCC. IEEE Solid-State Circuits Magazine. 2018;**10**(1):30-46

[29] Oboril F, Bishnoi R, Ebrahimi M, Tahoori MB. Evaluation of hybrid memory technologies using SOT-MRAM for on-chip cache hierarchy. IEEE Transactions on Computer-Aided Design of Integrated Circuits and Systems. 2015;**34**(3):367-380

[30] Seo Y, Kwon K, Fong X, Roy K. High performance and energy-efficient on-chip cache using dual port (1R/1W) spin-orbit torque MRAM. IEEE Journal on Emerging and Selected Topics in Circuits and Systems. 2016;**6**(3): 293-304

Chapter 9

Computationally Efficient Hybrid Interpolation and Baseline Restoration of the Brain-PET Pulses

Saeed Mian Qaisar

Abstract

The design and component level architectures of a novel offset compensated digital baseline restorer (BLR) and an original hybrid interpolator are described. It allows diminishing the effect of modifications occurring during the readout of Positron Emission Tomography (PET) pulses. Without treatment, such artifacts can result in a reduction in the scanner's performance, such as its sensitivity and resolution. The BLR recompenses the offset of PET pulses. Afterward, the pertinent parts of these pulses are located. Onward, the located portion of the signal is resampled by using a hybrid interpolator. This is constructed by cascading an optimized weighted least-square interpolator (WLSI) and a Simplified Linear Interpolator (SLI). The regulation processes for the WLSI coefficients and evaluation of the BLR and the interpolator modules are presented. The proposed hybrid interpolator's computational complexity is compared with classic counterparts. These modules are implemented in Very High-Speed Integrated Circuits Hardware Description Language (VHDL) and synthesized on a Field Programmable Gate Array (FPGA). The functionality of the system is validated with an experimental setup. Results reveal notable computational gain along with adequate dynamic restitution of the bipolar offsets besides a useful and accurate improvement of the temporal resolution relative to the computationally complex conventional equivalents.

Keywords: PET scanner, hybrid interpolation, baseline restoration, VHDL, FPGA, computational complexity, photo-detector, scintillator

1. Introduction

The recent technological developments in the field of microelectronics have revolutionized the development and deployment of biomedical implants, mobile healthcare, and biomedical scanners. In this framework, a variety of high-performance PET scanners have been proposed and realized during the last decade [1–3]. PET imaging is evolving the overall influence of nuclear medicine [1]. It is because of its superior performance, in terms of resolution, compared to the Single Photon Emission Computed Tomography (SPECT). Additionally, it is the rapidly

IntechOpen

growing nuclear diagnostic field [1]. It is a valuable metabolic imaging approach that is probably using the most suitable radiopharmaceutical [1].

Moreover, PET renders high-quality tomographs, which cannot be attained by SPECT or non-nuclear counterparts [1–3]. It has a significant clinical impact, which is evident from a variety of well-conducted case studies. Oncologists, using PET imaging services, appreciate its usefulness. PET is used in a number of critical clinical applications, including cancer diagnosis and monitoring, management of cardiology and cardiac surgery, and neurology and psychiatry.

Brain-PET scanner is an imaging device used for analysis and observation of the brain metabolic activity [1–3]. The concept is to inject a controlled volume of the radioactive tracer into the patient. Brain-PET frequently uses the tracer fluorine-18-fluorodeoxyglucose (18F-FDG) [4, 5]. After a certain period, a significant amount of this tracer is accumulated across the tumor cells. Tracer releases $\beta +$ positrons, annihilated by medium electrons. As a result of each annihilation, two γ-rays of 511-keV energies are emitted. These are released at about 180° relative angles at the same time. Crystal scintillators absorb energy from emitted γ-rays [6, 7]. They are turning the energy into photons of light. In the next step, photo-detectors pick up these photons, which transform them into electrical pulses [8–13].

The shape and amplitude of electrical pulses produced by photo-detectors enable undesirable interactions to be reduced [2, 3, 11, 14]. In this context, sophisticated PET pulses readout and conditioning are used to maximally preserve this information [12–15].

The PET pulses, produced by photo-detectors, are conditioned and processed to extract the information such as energies and timestamps. The extracted parameters are onward passed to the Image Reconstruction Module (IRM). It combines all pairs of γ-rays, produced by same annihilations, to generate the Line-of-Responses (LORs) [11, 15, 16]. It also uses energies of the PET pulses to measure the Depths of Interactions (DOIs) [1–3]. Afterward, IMR treats these LORs and DOIs to produce three-dimensional tomographs.

PET pulses readout introduces artifacts [11, 12, 16]. It degrades the exactitude of measuring the PET pulses energies and timestamps [11]. It influences device consistency in calculating DOIs and LORs, which lowers the scanner's sensitivity and resolution [11, 12, 16]. BLRs, pile-up correctors, and high-precision timestamp calculators are being used to resolve these deficits [11, 16–21].

The readout of photo-detectors' generated PET pulses produces a shift in the baseline of the signal. It diminishes the energy resolution and the precession of DOI calculation. BLRs are utilized in this framework. Various BLRs have been suggested, ranging from analog approaches to adaptive solutions based on digital filters [11, 12, 16–21].

Another primary method in PET scanners is the measurement of annihilation timestamps. It allows the collection of adequate annihilations to create LORs and also prevents the processing of ineffective information at IRM [11, 12]. Various timestamp calculators were suggested for this purpose. These are essentially formed by using discriminators and time to digital converters (TDCS) [11, 22–27].

The timestamp calculators based on analog discriminators do not face a problem with the temporal resolution of the input signal. This problem arises, however, in the case of timestamp calculators based on digital discriminators [11, 23]. The performance of the system relates directly to the temporal resolution of the input signal and the discriminatory algorithm. The interpolation and multi-threshold discriminators are used to achieve accurate measurement of digital timestamps [11, 28].

The remaining chapter is set out as follows. Section 2 explains the materials and methods. The VHDL-based system implementation and synthesis on FPGA are discussed in Section 3. Section 4 presents the experimental results. Section 5 makes a discussion and concludes the chapter.

2. Materials and methods

Figure 1 displays a block diagram of the proposed system. It illustrates that in the intended patient body "1" an appropriately controlled quantity of the radioactive tracer "2" is injected. The radioactive tracer used in this analysis is 18F-FDG. Most of this tracer is distributed through contaminated brain cells after a certain time [5]. It originates β+, which is annihilated with electrons of medium. Every annihilation emits two 511-keV energy γ-rays. These are simultaneously released at about 180° relative angle and interact with crystal scintillators of the detection sensors "4." Scintillators convert energies of γ-rays into photons of light. Afterward, photo-detectors pick up these photons and turn them into electrical pulses [8–13]. A determined number of sensors are used in a detection ring "3." The front-end electronics, located in the detection sensors "4," further prepare and process PET pulses emanating from photo-detectors. For the construction of tomographs, the selected pulses with extracted parameters like addresses of the intended crystal and the involved sensor are conveyed to the IRM "5."

2.1 The detection ring

The detection ring consists of a group of four sensors, arranged axially around an "B" axis in a circle. Most contemporary PET scanners are built from radially arranged scintillators [15, 23]. The use of scintillator axial arrangements with the

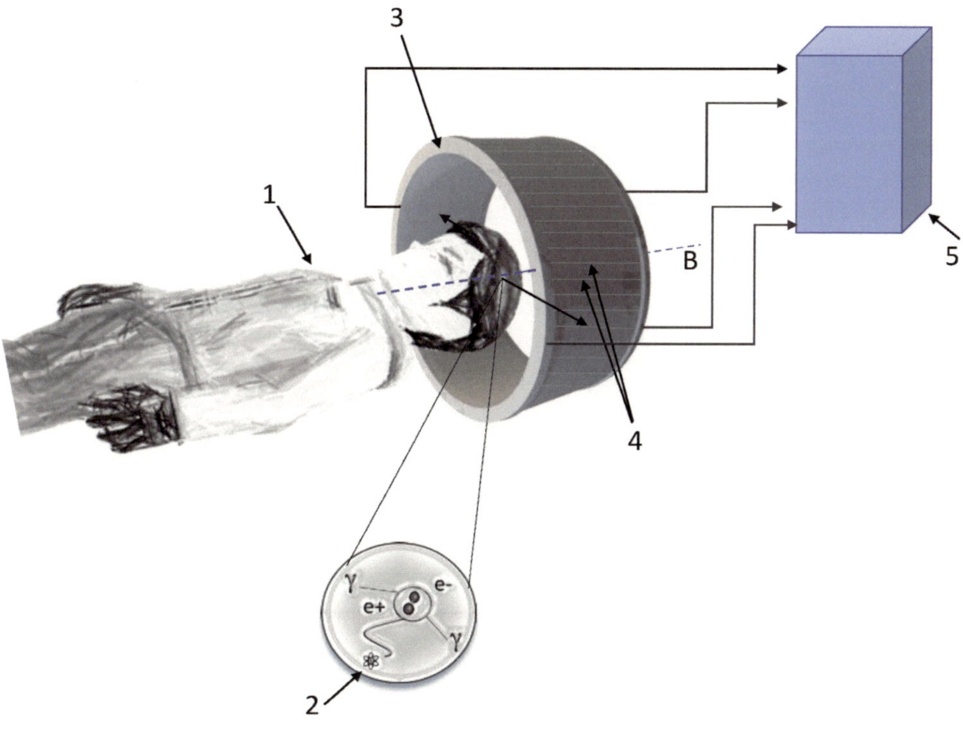

Figure 1.
Block diagram of the proposed system.

appropriate treatment of lateral sides of scintillators could increase the scanner's spatial resolution compared to alternatives, based on the radial arrangement of scintillators [11, 23].

Each sensor is composed of a scintillator crystal matrix. The LYSO scintillators are used in this study [6, 7]. The γ-rays interact with these crystals and turn their energies into photons of light. These photons are then sensed by two photo-detector matrices, positioned on both sides of the scintillator matrix. Such photo-detectors produce two electrical pulses as a result of each interaction [11, 26]. Each face of a crystal scintillator is positioned with a photo-detector. Only 2xP photo-detectors are used for a matrix of P crystal scintillators (cf. **Figure 2**). It reveals a γ-ray's interaction with one scintillator crystal "6." The crystal has faces "6a" and "6b" that are paired with two "7a" and "7b" MPPCs, respectively. The PET pulses produced by MPPCs are transferred to the electronics front-end module "8." These pulses are sorted and processed and the selected pulses with extracted information are conveyed to IRM "5."

A higher count of photons is attainable by MPPCs. Moreover, their behavior is free of the magnetic field influence. It is the reason behind the frequent usage of MPPCs in PET scanners as photo-detectors. The pixel size of the utilized MPPCs is $50 \times 50 \mu m^2$. Each MPPC contains 3600 pixels. The MPPC-generated pulses are of extremely low amplitude [20]. Hundreds of microamperes bound the highest amplitude [12]. Thus, they are noise sensitive and can be completely malformed. Such pulses are amplified to enhance their immunity. It is realized by cascading a charge resistance with an appropriate bandwidth amplifier [29]. Such amplified pulses are transferred to the blocks embedded in "8" for signal selection and address encoding. They pick pulses among a matrix for the active crystal and also ensure that only one crystal is enabled at a time. If this requirement is satisfied, then the further processing of the selected pair of pulses is realized.

An appropriate baseline for the selected pulses should be available for the correct functionality of the suggested BLR [12]. Two lumped delay lines are used for each sensor in this context [30]. They incorporate a delay of 50 ns, in the selected PET pulses, with a standard deviation of only about 100 ps.

The delayed pulses, after A/D conversion, are conveyed to the IRM. To attain appropriate accuracy, the sampling frequency, FS, and the quantizer resolution should be selected tactfully [11, 12, 29]. In this case, the selected and prepared PET pulses are acquired with two 12-bit resolution Analog-to-Digital Converters (ADCs), functioning at FS = 200 MHz. Besides the chosen pulses, addresses of the effective crystal and the sensor in question are also encoded and communicated to the IRM. These data, obtained from different sensors in the detection ring, are used by the IRM for calculating DOIs and LORs. In this way, the three-dimensional tomographs are produced.

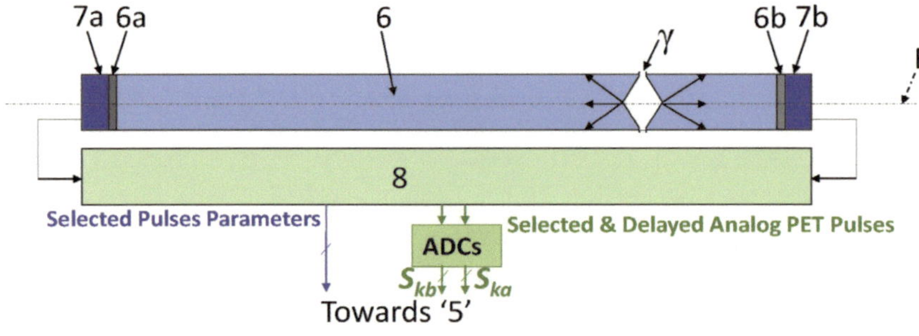

Figure 2.
The sensor principle.

2.2 Digital conditioning of the PET pulses

The amplification enhances the resistance against noise in the PET pulses [29]. Besides, this amplification can also inject random offsets in the processed pulses. It influences the performance of energy estimators and degrades DOI accuracy. In the same manner, the digitization of selected PET pulses degrades the temporal resolution. The reduced temporal resolution lowers the performance of the post timestamp and LORs' estimators. A rise in *FS* can enhance the temporal precision. It does, however, lead to a costly solution in terms of cost and power utilization [30].

The suggested BLR and interpolator processes concepts are illustrated by using **Figure 3**. It indicates that the BLR mainly processes the digitized versions of selected pulses in order to recover their baselines. Outputs of BLRs are conveyed to energy estimators and an adder. The added pulses are interpolated. It is performed to ameliorate the temporal resolution. Onward, the up-sampled signal is used to estimate the timestamp [11, 23].

2.3 The BLR

The BLR concept is shown in **Figure 4**. Two equivalent modules are introduced, working independently to retrieve the selected pair of PET pulses: S_{ka} and S_{kb} (cf. **Figure 3**).

Figure 4 shows that a real-time offset value, O_{Calc}, is first estimated from the incoming pulse by the BLR. It is determined as a mean of the baseline of the incoming pulse. The concept can be interpreted in mathematical terms using Eq. (1), where *N* represents the count of concerned samples, belonging to the digitized pulse baseline. S_{k_n} represents the sampled signal and *n* indexes the considered samples. *N* belongs to the set {1, 2, ..., *N*}. *N* is selected in accordance to the

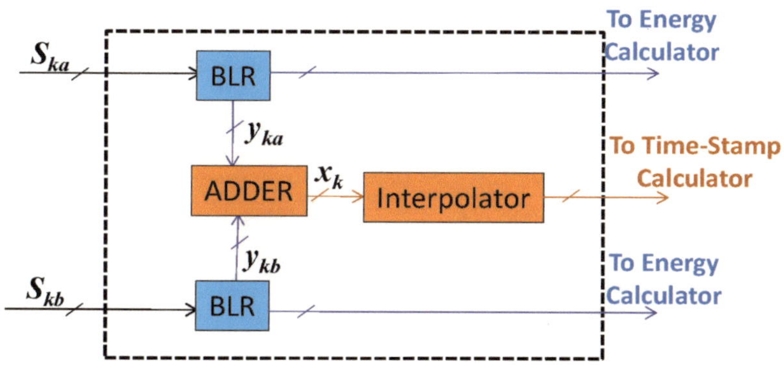

Figure 3.
The BLR and interpolator blocks.

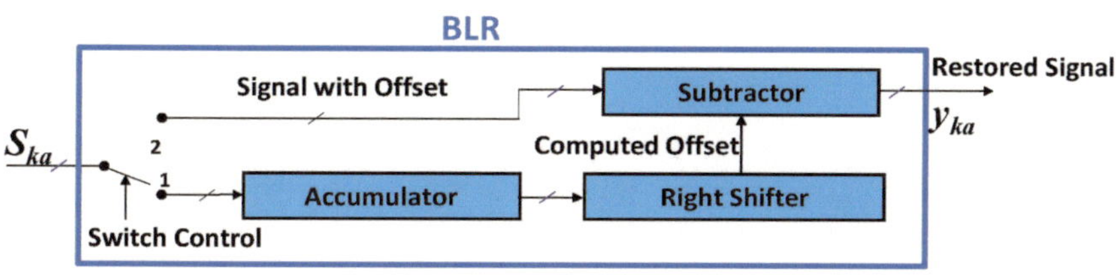

Figure 4.
The architecture of BLR module.

employed delay line and FS [29]. $N = 8$ is selected. It enables the mean value of the incoming pulse baseline to be estimated through accumulation and the right shifting while eviting the utilization of a complex conventional divider.

$$O_{Calc} = \frac{1}{N}\sum_{n=1}^{N} x_n \tag{1}$$

The restored pulse, free of offset, is obtained by employing Eq. (2) where y_{k_n} is the nth restored signal sample and S_{k_n} is the nth input signal sample.

$$y_{k_n} = S_{k_n} - O_{Calc} \tag{2}$$

This O_{Calc} estimation in real-time makes the suggested BLR self-adjustable. The value of O_{Calc} adapts as a function of the intended PET pulse. It allows an effective restoration of the incoming pulses with a diverse range of bipolar offsets [29].

2.4 The hybrid interpolator

The concept of the suggested hybrid interpolator is clear from the block diagram, shown in **Figure 5**.

2.4.1 The leading-edge selection

Figure 5 shows that the intended signal x_{k_n}, generated as sum of the outputs of both BLRs, is preliminary conveyed to the signal leading-edge selector. The attention on the leading-edge is due to the form of post timestamper, which is formed by combining multi-thresholds leading-edge discriminators [11].

The signal selection mechanism is illustrated by using **Figure 6**, where V_{max} is the input signal maximum amplitude, α is a percentage of V_{max} and is chosen equal to 10% of V_{max}. If the nth input signal sample, x_{k_n}, crosses the $\alpha.V_{max}$ threshold then $Q + 2$ samples are selected. The association among the nth nominated signal sample, xs_n, and the nth input signal sample, x_{k_n}, can be presented in mathematical terms with Eq (3).

$$xs_n = x_{k_{n+i}}, where\ i = \{-1, 0, 1, 2, .., N\} \tag{3}$$

This process of selection prevents processing of the entire signal length and thus dramatically improves the performance of the proposed system in terms of computation and power consumption [11].

The signal selector composes of a magnitude comparator, a circular buffer, and a module for logic and control. The concept is illustrated by using **Figure 7**. It displays that each input signal sample is compared with the predefined threshold $\alpha.Vmax$.

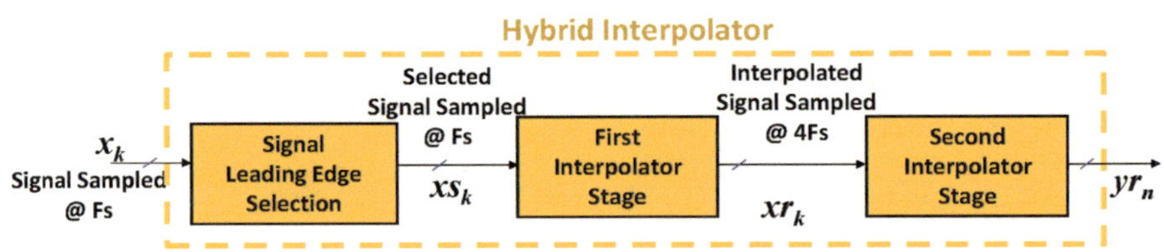

Figure 5.
The block diagram of hybrid interpolator block.

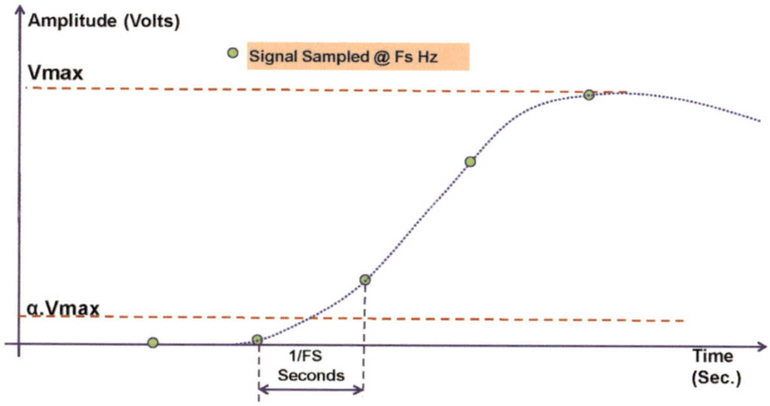

Figure 6.
Concept of signal selection.

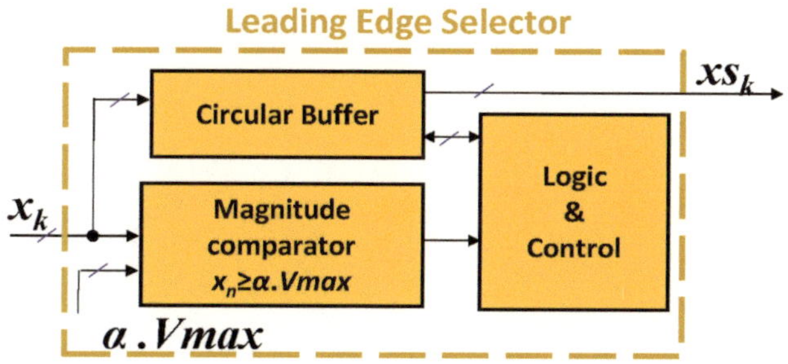

Figure 7.
The architecture of signal leading-edge selector.

Once that threshold is exceeded by an input sample, the magnitude comparator output will be high. It is used as a logic and control unit notification that allows the circular buffer to output the xs_n. The logic and control unit is based on a counter and a J-K latch. The comparator pilots the latch. Once set, it enables the output port of the buffer. The counter provides the address of buffer registers needed to read. Finally, the logic and control unit resets the J-K latch, after reading the xs_n.

2.4.2 First stage interpolator

The chosen portions of pulses are up-sampled with interpolation factor, $IF = 4$, by utilizing the primary interpolator stage. It means three equally spaced samples are positioned statistically between the initial two consecutive samples. It fourfold ameliorates the first stage output temporal resolution compared to its input. The Weighted Least Square Interpolator (WLSI) up-samples at the first stage [11].

WLSI coefficients are determined using 10,000 summed pulses. The summed pulses provided by the adder "9" are used as reference ones (cf. **Figure 9**). It allows the interpolated values to be compared with the reference ones and adapts the interpolation coefficients to reduce the differences from the reference signal. In particular, these coefficients are determined by utilizing the Least Square (LS) algorithm to diminish the squares of discrepancies among the interpolated values and the actual ones. Five samples of xs_n are passed to the first stage interpolator input (cf. Section 3). After up-sampling, it outputs 17 samples. The error functions of the WLSI coefficients for the 12 implanted samples are estimated as:

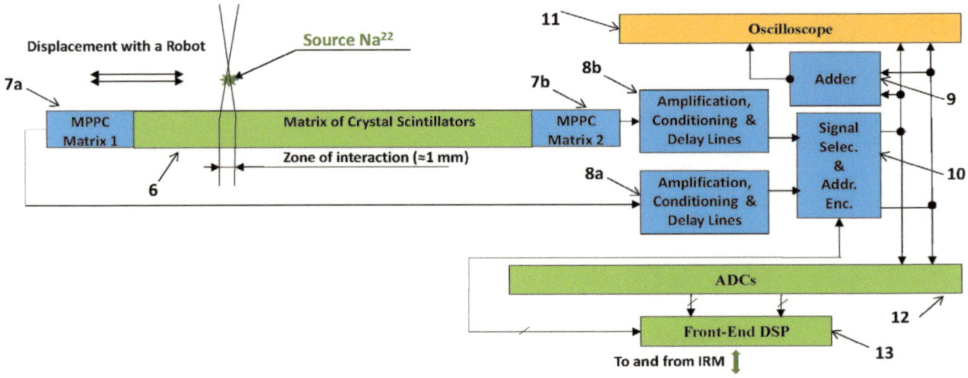

Figure 8.
The experimental setup.

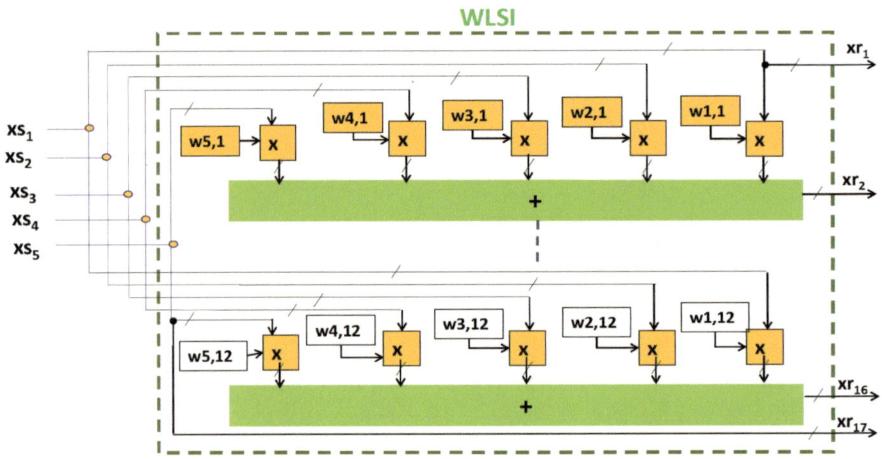

Figure 9.
The architecture of WLSI.

$$ferr_1(w_{1,1}, w_{2,1}, \dots w_{5,1}) = \sum_{Np=1}^{K} \sum_{n=1}^{C} \left(xs_1.w_{1,1} + xs_2.w_{2,1} + \dots xs_5.w_{5,1} - xrefNp_n\right)^2. \tag{4}$$

$$ferr_2(w_{1,2}, w_{2,2}, \dots w_{5,2}) = \sum_{Np=1}^{K} \sum_{n=1}^{C} \left(xs_1.w_{1,2} + xs_2.w_{2,2} + \dots xs_5.w_{5,2} - xrefNp_n\right)^2. \tag{5}$$

$$\vdots$$

$$ferr_{12}(w_{1,12}, w_{2,12}, \dots w_{5,12}) = \sum_{Np=1}^{K} \sum_{n=1}^{C} \left(xs_1.w_{1,12} + xs_2.w_{2,12} + \dots xs_5.w_{5,12} - xrefNp_n\right)^2 \tag{6}$$

In Eqs. (4)–(6), K shows the total number of summed pulses, used in the measurement. In this case, $K = 5000$ is selected. C is the count of samples per intended pulse. Eq. (7) determines the matrix of coefficients w, referring to the minimum errors.

$$w = \left(X^T.X\right)^{-1}.X^T.X_{ref} \tag{7}$$

where X is the matrix of the initial values used for the interpolation. $Xref$ is the matrix of reference samples at the time instants where approximations are made.

$$X = \begin{pmatrix} xs1_1 & \dots & xs1_5 \\ \dots & \dots & \dots \\ xsNp_1 & \dots & xsNp_5 \end{pmatrix} \tag{8}$$

$$Xref = \begin{pmatrix} xref1_1 & \dots & xref1_{12} \\ \dots & \dots & \dots \\ xrefNp_1 & \dots & xrefNp_{12} \end{pmatrix} \tag{9}$$

The program for setting the X and $Xref$ matrices and carrying out the matrix computations to calculate the WLSI coefficients is specifically implemented in MATLAB [31]. The architecture utilized by the designed WLSI can be seen in **Figure 10**.

2.4.3 Second stage interpolator

The output of first stage is further up-sampled by the second stage interpolator. It is presumed that the signal between two successive samples changes linearly. Therefore, in the second stage, the SLI with an IF = 4 is deployed. The value of nth approximated value, xr_n, corresponding to the nth interpolating instant, tr_n, is equal to the mean of its previous and following incoming samples (cf. Eq. (10)).

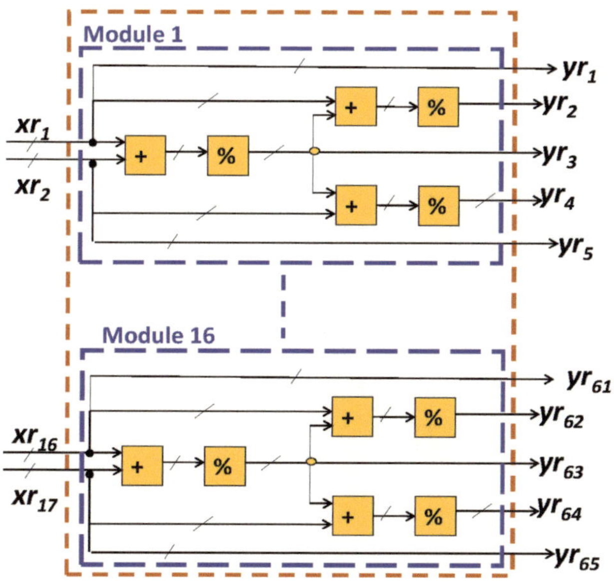

Figure 10.
The architecture of SLI.

$$xr_n = \frac{xs_{n-1} + xs_n}{2} \tag{10}$$

In this way, the temporal resolution of the second stage interpolator output is 16-fold superior than that of the original incoming signal. The second stage interpolator collects 17 samples from the prior WLSI module and outputs 65 samples. **Figure 8** illustrates the architecture and working of SLI. It indicates that 16 identical modules are operating concurrently.

2.5 The experimental setup

The system applicability is studied by using an experimental setup, as shown in **Figure 9**. It indicates that the radioactive tracer, ^{22}Na, is relocated by a robotic arm with respect to the utilized crystal scintillators matrix "6." The gap among successive relocation steps remains equivalent to 1 mm. The system includes a sensor, consisting of four LYSO crystals of $3 \times 3 \times 60$ mm^3, enclosed from both sides with 3×3 mm^2 arrays of Hamamatsu MPPCs, "7a" and "7b." MPPC array consists of four MPPCs of 3×3 mm^2 area, to suit the used scintillators.

The front-end electronics modules, "8a" and "8b," manage the pulses, which come from both MPPC matrices "7a" and "7b." They realize amplification, selection, and addition of delay. Signal selection and address encoding block "10" is used to process outputs of "8a" and "8b." The selected pulses are then added using the adder circuit "9." Finally, the selected pulses supplied by "10" and their sum, supplied by "9" are digitized concurrently by using an oscilloscope. The LeCroy WavePro 7300A oscilloscope is utilized in this study. The pulses, produced by "10" are digitized at FS = 200 MHz. The sum of pulses produced by "9" is digitized at the sampling rate of 3.2GHz. Later, when characterizing the hybrid interpolation module, these pulses, obtained at a frequency of 3.2 GHz, are used as reference ones. The approach is described further in Section 3.2.

3. VHDL implementation and synthesis on FPGA

The designed BLR and hybrid interpolator modules are implemented in VHDL and are synthesized on an FPGA. Preliminarily the 28-nm technology-based "xc7a200t FPGA" from the "Artix-7" family is considered [32]. The circuit VHDL implementation is synthesized on the selected FPGA by using the Xilinx Synthesis Technology (XST) [33, 34]. The system Register-Transfer-Level (RTL) schematics are shown in **Figure 11a** and **Figure 11b**. **Figure 11a** shows the top-level RTL schematic and **Figure 11b** shows the components-level RTL schematic.

The post-synthesis summary of the FPGA resource utilization, by one BLR and hybrid interpolator, is presented in **Table 1**. It shows that more than 10 proposed BLR and hybrid interpolator chains can be implemented on a single *xc7a200t* chip. The available DSP-blocks, on a single *xc7a200t* chip, mainly pose this limitation on the count of proposed conditioning chains. In case of need, this limitation can be resolved by using alternate logic-cells-based architectures and implementations, which minimize the use of DSP-blocks.

4. Results

The statistical features of the PET pulses such as rise-time, fall-time, bandwidth, and magnitude are extracted. It is carried out by analyzing the PET pulses delivered

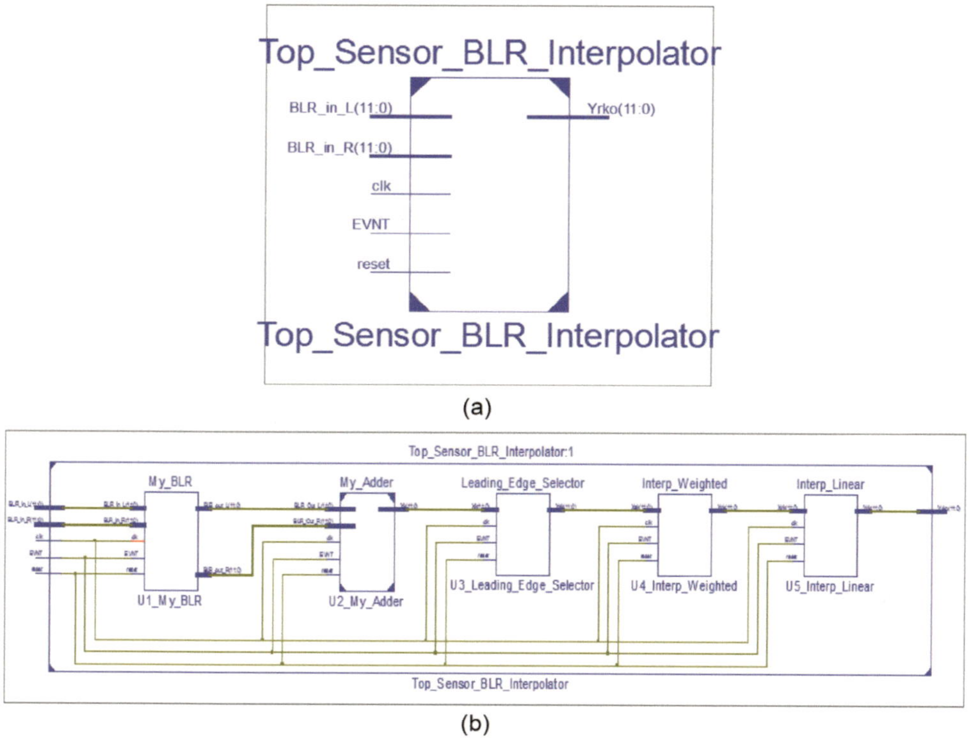

Figure 11.
(a) The top-level RTL schematic. (b) The components-level RTL schematic.

Resource	Input-output pins (IOs)	Slice registers	Slice LUTs	DSP48E1s	Block RAMs
Used	36	1223	818	60	8
Total	500	269,200	134,600	740	365
% utilization	7.2	0.5	0.6	8.1	2.2

Table 1.
Summary of the xc7a200t resources utilization.

Parameter	Mean value	Standard deviation
Rise-time (ns) 10–90%	11	1.1
Fall-time (ns) 10–90%	146	10
Bandwidth (MHz)	35	4
Maximum amplitude, Vmax, after addition of both pulses(mV)	1010	55

Table 2.
Summary of amplified and conditioned pulses parameters.

by "10." Ten thousand pulses are considered through this operation. They are collected at a sampling rate of 200 MHz by utilizing the inbuilt ADCs of the oscilloscope. A summary is presented in **Table 2**.

4.1 The characterization of BLR

The utilized delay lines inject 50-ns delays in the selected pulses. These are acquired with a sampling rate of 200 MHz. Thus, 10 samples are collected on the baselines of arriving pulses. While estimating O_{Calc}, $N = 8$ is selected. It refers to primarily 40-ns portion of the baseline. On the accumulator output, the division with a factor of 8 is attained by executing a 3-bit right shift.

Figure 12 shows a typical performance of the conceived BLR. It depicts how well the offset, injected by the front-end electronics, is compensated from the received pulses. Amplification and conditioning chains in modules "8a" and "8b" add a particular offset in the received pulses. Due to the influence of various offsets, for a fixed location among the radiotracer and the crystal scintillator matrix "6," the incoming pulses demonstrate a peak amplitude dispersion of roughly 25% among them. With the application of the designed BLR, these dispersions are limited to approximately 1%. It reflects how the use of designed BLR will enhance the correctness of post energies and DOI estimators. It will also improve the performance of the scanner in terms of precision in the localization of tumor cells [11, 12].

4.2 The characterization of hybrid interpolator

4.2.1 The leading-edge selector

BLRs' outputs are summed and then transferred to the signal selection unit. The selected pulses are digitized at a rate of 200 MHz. **Table 1** reveals that the incoming signal's mean [10] % rise time is 11 ns. It clarifies that there will be at least two samples, in the digitized version of the signal, on the signal leading-edge. Hence, $Q = 3$ is selected. It ensures a proper selection of the signal leading-edge, the most critical signal portion needed by the used timestamp estimator [11].

4.2.2 The first and the second stage interpolators

Signal leading-edge selector output is transferred to the first interpolation stage realized with WLSI. In this study, this interpolation stage receives five selected signal samples, initially sampled at 200 MHz. It outputs 17 samples, with a sampling rate of 800 MHz.

The SLI is selected to process the second interpolation stage. It receives 17 samples and outputs 65 samples with a sampling rate of 3.2 GHz.

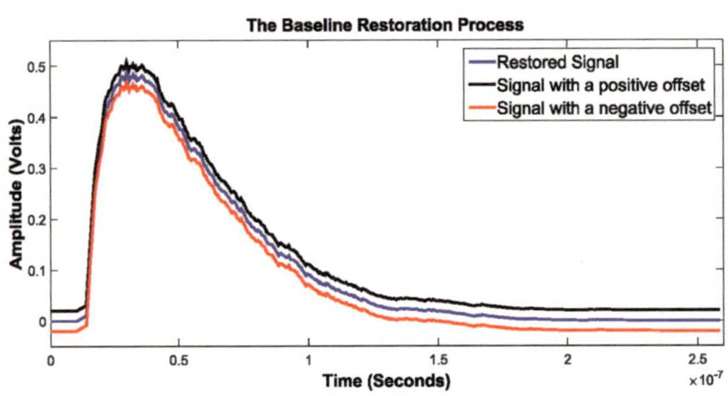

Figure 12.
The baseline restoration process.

The interpolation errors are computed. Throughout this procedure, 10,000 interactions among the radioactive source and the scintillators matrix "6" are used. The intended interactions are recorded at predetermined locations.

The module "10" produces a pair of pulses as an outcome of each interaction. These pulses and their related sum are digitized with an oscilloscope. The pulses produced by the module "10" are recorded at a sampling frequency of 200 MHz. The sum of pulses produced by the adder circuit "9" is recorded at a sampling rate of 3.2 GHz. The error per interpolated sample, Ie_n, is estimated with Eq. (11), where, $xref_n$ is the reference sample value with respect to the interpolation instant tr_n. It represents the summed analog signal that is sampled with an oscilloscope, at 3.2 GHz. yr_n is the approximated value computed with respect to the interpolation instant tr_n.

$$Ie_n = xref_n - yr_n \tag{11}$$

The error of the used interpolator is measured in terms of the Mean Square Deviation Error (RMSDE). The RMSDE is determined for each intended pulse by using Eq. (12), where, C is the number of samples that are considered for the incoming pulse. The number of summed pulses is indexed by Np. Eventually, the RMSDE mean is determined as the average value of the $RMSDE_{Np}$.

$$RMSDE_{Np} = \sqrt{\frac{1}{C} \cdot \sum_{n=1}^{C} Ie_n^2} \tag{12}$$

For designed hybrid interpolator, the average RMSDE value is 13.6 μV. The interpolated signal is used for post timestamps estimation. The interaction time among the γ-ray and the crystal scintillator is computed by comparing the arriving pulse amplitude against determined thresholds. The time is estimated on the basis of thresholds crossing instants. Due to the discrete-time digitized pulses, a threshold crossing is estimated as intersection among the threshold and the straight line, crossing via two successive samples that lay across that threshold. For the timestamp estimators, based on digital discriminators, the precession of timestamps is directly related to the temporal resolution and the magnitude accuracy of the arriving pulses [11, 23].

In this case, by using the suggested hybrid interpolator, the temporal resolution of the selected portion is enhanced 16-fold. It outputs an interpolated signal with 0.3125 ns of temporal resolution and 13.6 μV of average RMSDE. It is capable of significantly improving the accuracy of the post timestamp estimator while digitizing pulses with ADCs that are economically accessible [11].

4.3 Comparison of the hybrid interpolator with a conventional interpolator

The effectiveness of the suggested hybrid interpolator is also compared with a counter mono-algorithm-based approach. Interpolation is achieved traditionally by using a unique algorithm. The computational load is directly related to the order of the interpolator and to the count of input samples.

The leading-edge selection module, in the proposed solution, allows concentrating on the pertinent portion of the signal. In the examined case, five samples per incoming pulse are picked. Without this element, however, the entire pulse duration of around 150 ns should be considered [11, 12]. The ADC will deliver 30 samples per incoming pulse, for a sampling frequency of 200 MHz. It will increase the interpolator computational load with a factor of 6.

The computational complexity for the suggested case is determined by adding the computational costs of both stages. The computational complexity of the optimized WLSI with an *IF* = 4 is 15 multiplications and 3 additions (cf. **Figure 10**). The computational complexity of SLI is 3 additions and 3 binary-weighted divisions (cf. **Figure 8**). Relative to the addition and multiplication operations, the circuit level complexity of the binary weighted division is insignificant. However, the complexity of a WLSI with an IF = 16 is 45 multiplications and 15 additions for approximating 15 samples among two consecutive originals [12].

By the grace of leading-edge selector, in the proposed case, the WLSI needs to interpolate 5 selected samples and it delivers 17 interpolated samples at its output. The overall computational cost of WLSI becomes 75 multiplications and 15 additions. The complexity of used SLI for processing 17 samples is 51 additions. It results in an overall cost of 75 multiplications and 66 additions. Contrary, in conventional case, the system has to process 30 samples for the whole pulse length, between 10% rise-time to 10% fall-time, of around 150 ns should be considered [1, 2]. It results in a computational cost of 1350 multiplications and 450 additions.

These results demonstrate that the suggested solution reaches an 18-fold gain in the count of multiplications and a 6.8-fold gain in the count of additions over the mono WLSI based solution.

The proposed hybrid interpolator's interpolation precision is also contrasted with the mono WLSI-based solution. Characterization is achieved by using the same pair of pulses registered and used during the measurement of the hybrid interpolator error. The interpolation error of the up-sampled signal outputs by the mono WLSI interpolator is estimated by using Eq. 12. It results in 11.4-μV RMSDE. For the case of designed hybrid interpolator, 13.6-μV RMSDE is attained. It reveals that the designed solution achieves substantial computational advantage over the mono WLSI-based tactic while achieving an analogous precision.

5. Discussion and conclusion

A brain-PET scanner's resolution and sensitivity rely on how accurate it is to measure the depths and times of interactions among scintillators and the γ-rays [1–3]. The exact measurement of the DOI requires an accurate estimate of the arriving PET pulses' energies [12]. A precise measurement of interaction times needs an exact calculation of timestamps [11]. This helps in accurate reconstruction of the LORs. Accurate DOIs and LORs contribute to the reconstruction of three-dimensional high-resolution tomography, which allows a precise location of tumor cells in the patient's brain.

The readout electronics of PET pulses produce unpredictable offsets. It causes an incorrect estimation of these pulses' energies, which decreases the accuracy of measurement of the DOI and timestamp. As a result, it diminishes the scanner's ability to assess the location of tumor cells accurately. An efficient BLR is proposed for minimizing the impact of offsets. It attains a dynamic offset cancelation by employing a real-time *OCalc* estimation mechanism. By the grace of the suggested BLR, the peak amplitude dispersions for a determined position of the radioactive source with respect to the matrix of scintillators "6" is diminished from ≤25% to ≤1%. This assures that the use of designed BLR would enhance the precision of post energies and DOIs' estimators. This should improve the scanner's accuracy in the localization of tumor cells [1–3, 11].

After energies, the second significant parameter to be measured in brain-PET scanners is the timestamp of the annihilated γ-rays [11]. It is conducted to calculate LORs, which allow the tumor cells to be located. A PET scanner's sensitivity and tomographic resolution are directly linked to the computational accuracy of timestamps [11, 23]. The timestamp can be measured in either digital or analog worlds. The analog timestamp calculators require the development of complex, integrated circuits for specific applications. Using digital timestamp calculators can result in a cost-effective solution [23]. It allows a solution to be realized using regular ADCs and Field Programmable Gate Arrays (FPGAs). Comparing the amplitude of digitized versions of PET pulses to established thresholds, they measure timestamps [23, 24, 35]. The time is stamped by using the instants of the threshold crossing. Hence, the exactitude of timestamp calculations is directly related to the incoming pulses' temporal resolution and precision of magnitude. The incoming pulses are digitized at a sampling rate of 200 MHz in the examined case. It provides 5 ns of temporal resolution. Later, with the suggested hybrid interpolator, the temporal resolution of the selected portion of pulses is 16-fold enhanced. It results in an interpolated signal with a temporal resolution of 0.3125 ns and with RMSDE of 13.6 μV. It aptitudes a significant improvement in the precision of the post timestamp calculator while acquiring the pulses with economically available ADCs.

Component-level architectures of the suggested BLR and hybrid interpolator are described. The proposed chain is implemented in VHDL, and synthesis is realized on the *xc7a200t* FPGA. It is shown that more than 10 proposed BLR and hybrid interpolator chains can be implemented on a single *xc7a200t* chip, which costs around 260US$. This reveals that the suggested concept can be developed, unlike traditional predecessors, by using cost-effective ADCs and FPGAs [12]. It prevents the production of complex high-performance specific integrated circuits and thus results in effective realization. In addition to cost-effectiveness, it also facilitates the device reconfiguration compared to hardwired circuits and allows similar precision to be attained [12].

These results demonstrate the potential applicability of the proposed BLR and hybrid interpolator in current brain-PET scanners. They can be easily incorporated into contemporary PET scanners based on a simple architecture and can contribute effectively in improving their tomographic resolution.

Acknowledgements

The author is thankful to Dr. D. Liksonov, Dr. A. Bakkali, and Dr. N. Tamda for the fruitful discussions, Mme. Rabia Qaisar, Ms. M. Tayeb, and Ms. E. Noor for their help in preparing figures for this book chapter.

Conflict of interest

The author declares no conflict of interest.

Author details

Saeed Mian Qaisar
Effat University, College of Engineering, Jeddah, Saudi Arabia

*Address all correspondence to: sqaisar@effatuniversity.edu.sa

References

[1] Fraioli F, Almansory KO. Clinical applications of PET/MRI in brain imaging. In: PET/CT in Brain Disorders. Cham: Springer; 2019. pp. 145-154. DOI: 10.1007/978-3-030-01523-7_13

[2] Ota R, Omura T, Yamada R, Miwa T, Watanabe M. Evaluation of a sub-millimeter resolution PET detector with a 1.2 mm pitch TSV-MPPC array one-to-one coupled to LFS scintillator crystals and inter-crystal scatter studies with individual signal readout. IEEE Transactions on Radiation and Plasma Medical Sciences. 2016;**1**(1):15-22. DOI: 10.1109/TNS.2016.2617334

[3] Gaudin É, Toussaint M, Thibaudeau C, Paillé M, Fontaine R, Lecomte R. Performance simulation of an ultrahigh resolution brain PET scanner using 1.2-mm pixel detectors. IEEE Transactions on Radiation and Plasma Medical Sciences. 2018;**3**(3): 334-342. DOI: 10.1109/TRPMS.2018.2877511

[4] Dell'Aquila AM, Avramovic N, Mastrobuoni S, Motekallemi A, Wisniewski K, Scherer M, et al. Fluorine-18 fluorodeoxyglucose positron emission tomography/computed tomography for improving diagnosis of infection in patients on CF-LVAD: Longing for more 'insights'. European Heart Journal Cardiovascular Imaging. 2018;**19**(5):532-543. DOI: 10.1093/ehjci/jex158

[5] Huang Q, Ren S, Jiang D, Guan Y, Xie F, Sun D, et al. Changes in brain glucose metabolism and connectivity in somatoform disorders: An 18 F-FDG PET study. European Archives of Psychiatry and Clinical Neuroscience. 2019:1-11. DOI: 10.1007/s00406-019-01083-0

[6] Marcinkowski R, Mollet P, Van Holen R, Vandenberghe S. Sub-millimetre DOI detector based on monolithic LYSO and digital SiPM for a dedicated small-animal PET system. Physics in Medicine & Biology. 2016; **61**(5):2196. DOI: 10.1088/0031-9155/61/5/2196

[7] Wolszczak W, Ereiksson L, Melcher CL. Scintillators for PET and SPECT. In: Physics of PET and SPECT Imaging. CRC Press; 2017. pp. 65-84. DOI: 10.4324/9781315374383

[8] Ronzhin A. High time-resolution photo-detectors for PET applications. Nuclear Instruments and Methods in Physics Research Section A: Accelerators, Spectrometers, Detectors and Associated Equipment. 2016;**809**:53-57. DOI: 10.1016/j.nima.2015.11.069

[9] Slawomir SP. A Technical Guide to Silicon Photomultipliers (SiPM) | Hamamatsu Photonics. Japan: Hamamatsu Corporation & New Jersey Institute of Technology; 2017

[10] Yamamoto S, Imaizumi M, Watabe T, Watabe H, Kanai Y, Shimosegawa E, et al. Development of a Si-PM-based high-resolution PET system for small animals. Physics in Medicine and Biology. 2010;**55**(19): 5817. DOI: 10.1088/0031-9155/55/19/013

[11] Qaisar SM. A two stage interpolator and multi threshold discriminator for the brain-PET scanner timestamp calculation. Nuclear Instruments and Methods in Physics Research Section A: Accelerators, Spectrometers, Detectors and Associated Equipment. 2019;**922**: 364-372. DOI: 10.1016/j.nima.2019.01.004

[12] Qaisar SM. Offset compensated baseline restoration and computationally efficient hybrid interpolation for the brain PET. Bio-Algorithms and Med-Systems. 2018; **14**(4). DOI: 10.1515/bams-2018-0031

[13] Qaisar SM. An automatic Peltier effect based MPPC gain regulation for brain-PET. In: 2017 International Conference on Electrical and Computing Technologies and Applications (ICECTA). UAE: IEEE; 2017. pp. 1-6. DOI: 10.1109/ICECTA.2017.8252028

[14] Qaisar SM. A VHDL based interpolator and base line restorer for a brain-PET scanner. In: 2017 International Conference on Electrical and Computing Technologies and Applications (ICECTA). UAE: IEEE; 2017. pp. 1-6. DOI: 10.1109/ICECTA.2017.8252056

[15] Moskal P, Bednarski T, Białas P, Czerwiński E, Kapłon Ł, Kochanowski A, et al. A novel method based solely on field programmable gate array (FPGA) units enabling measurement of time and charge of analog signals in positron emission tomography (PET). Bio-Algorithms and Med-Systems. 2014;**10**(1):41-45. DOI: 10.1515/bams-2013-0104

[16] Lu J, Zhao L, Chen K, Deng P, Li B, Liu S, et al. Real-time FPGA-based digital signal processing and correction for a small animal PET. IEEE Transactions on Nuclear Science. 2019; **66**(7):1287-1295. DOI: 10.1109/TNS.2019.2908220

[17] Saxena S, Hawari AI. Digital pulse deconvolution with adaptive shaping for real-time high-resolution high-throughput gamma spectroscopy. In: Nuclear Instruments and Methods in Physics Research Section A: Accelerators, Spectrometers, Detectors and Associated Equipment. Elsevier; 2018. DOI: 10.1016/j.nima.2018.09.123

[18] Hammad ME, Kasban H, Fikry RM, Dessoky MI, Zahran O, Elaraby SM, et al. Pile-up correction algorithm for high count rate gamma ray spectroscopy. Applied Radiation and Isotopes. 2019;**151**:196-206. DOI: 10.1016/j.apradiso.2019.06.003

[19] Kawada Y, Yunoki A, Yamada T, Hino Y. Effect of time walk in the use of single channel analyzer/discriminator for saturated pulses in the 4πβ–γ coincidence experiments. Applied Radiation and Isotopes. 2016;**109**: 369-373. DOI: 10.1016/j.apradiso.2015.12.024

[20] Du J, Schmall JP, Judenhofer MS, Di K, Yang Y, Cherry SR. A time-walk correction method for PET detectors based on leading edge discriminators. IEEE Transactions on Radiation and Plasma Medical Sciences. 2017;**1**(5): 385-390. DOI: 10.1109/TRPMS.2017.2726534

[21] Kafaee M, Moussavi-Zarandi A. Baseline restoration and pile-up correction based on bipolar cusp-like shaping for high-resolution radiation spectroscopy. Journal of the Korean Physical Society. 2016;**68**(8):960-964. DOI: 10.3938/jkps.68.960

[22] Moskal P, Zoń N, Bednarski T, Białas P, Czerwiński E, Gajos A, et al. A novel method for the line-of-response and time-of-flight reconstruction in TOF-PET detectors based on a library of synchronized model signals. Nuclear Instruments and Methods in Physics Research Section A: Accelerators, Spectrometers, Detectors and Associated Equipment. 2015;**775**:54-62. DOI: 10.1016/j.nima.2014.12.005

[23] Liksonov D, Qaisar SM. Method for Calculating Interaction Time of Gamma Photon with Scintillation Crystal in Sensor of Positron Emission Tomography System, Involves Determining Curve Representing Evolution of Electrical Signal by Interpolation of Sampling Points, FR 2983590. Available from: https://worldwide.espacenet.com/patent/search?q=pn%3DFR2983590B1

[24] Braga LH, Gasparini L, Grant L, Henderson RK, Massari N, Perenzoni M, et al. A fully digital 8x16 SiPM array for PET applications with per-pixel TDCs and real-time energy output. IEEE Journal of Solid-State Circuits. 2014; **49**(1):301-314

[25] Braga LH, Gasparini L, Stoppa D. A time of arrival estimator based on multiple timestamps for digital PET detectors. In: 2012 IEEE Nuclear Science Symposium and Medical Imaging Conference (NSS/MIC). California, USA: IEEE; 2012. pp. 1250-1252. DOI: 10.1109/NSSMIC.2012.6551306

[26] Tetrault MA, Oliver JF, Bergeron M, Lecomte R, Fontaine R. Real time coincidence detection engine for high count rate timestamp based PET. IEEE Transactions on Nuclear Science. 2010; **57**(1):117-124. DOI: 10.1109/TNS.2009.2038055

[27] Xie Q, Kao CM, Wang X, Guo N, Zhu C, Frisch H, et al. Potential advantages of digitally sampling scintillation pulses in timing determination in PET. In: 2007 IEEE Nuclear Science Symposium Conference Record (NSS'07). Vol. 6. Honolulu, Hawaii, USA: IEEE; 2007. pp. 4271-4274. DOI: 10.1109/NSSMIC.2007.4437060

[28] Pałka M, Strzempek P, Korcyl G, Bednarski T, Niedźwiecki S, Białas P, et al. Multichannel FPGA based MVT system for high precision time (20 ps RMS) and charge measurement. Journal of Instrumentation. 2017;**12**(08): P08001. DOI: 10.1088/1748-0221/12/08/P08001

[29] Qaisar SM. A custom 70-channel mixed signal ASIC for the brain-PET detectors signal readout and selection. Biomedical Physics & Engineering Express. 2019;**5**(4):045018. Available from: https://orcid.org/0000-0002-4268-3482

[30] Hanu AR, Prestwich WV, Byun SH. A data acquisition system for two-dimensional position sensitive micropattern gas detectors with delay-line readout. Nuclear Instruments and Methods in Physics Research Section A: Accelerators, Spectrometers, Detectors and Associated Equipment. 2015;**780**: 33-39. DOI: 10.1016/j.nima.2015.01.053

[31] Rao KD, Tunga S. Signals and Systems. Switzerland: Birkhäuser; 2018. DOI: 10.1016/C2009-0-19375-3

[32] Mohsen E. Reducing system power and cost with Artix-7 FPGAs. Xilinx, Artix. 2012;**7**:1-12

[33] Wang Y, Li P, Zhang P, Zhang C, Cong J. Memory partitioning for multidimensional arrays in high-level synthesis. In: Proceedings of the 50th Annual Design Automation Conference. New York, USA; 2013. pp. 1-8. DOI: 10.1145/2463209.2488748

[34] Agrawal T, Kumar A, Saraswat SK. Design of low power SRAM on Artix-7 FPGA. In: 2016 2nd International Conference on Communication Control and Intelligent Systems (CCIS). India: IEEE; 2016. pp. 203-209. DOI: 10.1109/CCIntelS.2016.7878231

[35] Li H, Wang C, Baghaei H, Zhang Y, Ramirez R, Liu S, et al. A new statistics-based online baseline restorer for a high count-rate fully digital system. IEEE Transactions on Nuclear Science. 2010; **57**(2):550-555. DOI: 10.1109/TNS.2009.2036914